Readings in
Farming Systems
Research and Development

Also of Interest

*Available in hardcover and paperback.

Westview Special Studies in Agriculture/Aquaculture Science and Policy

Readings in Farming Systems Research and Development
edited by W. W. Shaner, P. F. Philipp, and W. R. Schmehl

This collection offers a comprehensive view of the commonalities and diversities of the farming systems research and development (FSR&D) approaches being applied around the world. The authors--among the leading practitioners in FSR&D--discuss conceptual frameworks, research methodology, data collection, and several ongoing FSR&D programs. The book is a must for anyone interested in gaining a concise, yet broad view of this new and growing field of research and its importance to small-scale farming in developing countries.

Dr. Shaner, project director of the Farming Systems Research and Development Methodology Project and associate professor of engineering at Colorado State University, has been chief economic advisor to the Technical Agency, Ministry of Planning, Ethiopia, and associate director of the Consortium for International Development. He has written *Project Planning for Development* (1979) and coauthored with Professors Philipp and Schmehl *Farming Systems Research and Development: Guidelines for Developing Countries* (Westview, 1982). Dr. Philipp is professor emeritus in the Department of Agricultural and Resource Economics at the University of Hawaii. Dr. Schmehl is professor and associate department head of agronomy at Colorado State University.

A Consortium for International Development Study

Readings in Farming Systems Research and Development

edited by W. W. Shaner,
P. F. Philipp, and W. R. Schmehl

Routledge
Taylor & Francis Group

LONDON AND NEW YORK

First published 1982 by Westview Press

Published 2019 by Routledge
52 Vanderbilt Avenue, New York, NY 10017
2 Park Square, Milton Park, Abingdon, Oxon OX14 4RN

Routledge is an imprint of the Taylor & Francis Group, an informa business

Library of Congress Catalog Card Number 82-50803

ISBN 13: 978-0-367-28509-8 (hbk)
ISBN 13: 978-0-367-30055-5 (pbk)

Contents

viii

Figures and Tables

FIGURES

Preface

In October 1978, the United States Agency for International Development contracted with the Consortium for International Development (CID) to prepare a set of guidelines on farming systems research and development (FSR&D). In carrying out the contract, CID gave Colorado State University (CSU) lead responsibility and subcontracted portions of the work to the University of Hawaii. Based on that contract, a book of guidelines-- *Farming Systems Research and Development: Guidelines for Developing Countries*-- was written primarily for research and development institutions in the developing countries. Another product of that contract is this book of readings on FSR&D.

In researching the materials for our book of guidelines, we found that some scientists had conducted considerable research in FSR&D--primarily in cropping systems--and a few development groups had been successful in implementing the approach. However, much of this work was scattered throughout the world and published results had not been widely distributed. Consequently, one of our project team's first tasks was to contact institutions and individuals who were working in FSR&D or related areas. This initial reconnaissance culminated in a workshop in FSR&D held at CSU August 1-4, 1979. Some of the world's leading practitioners in FSR&D participated in this workshop.

The selected readings in this book contain papers prepared by these practitioners and illustrate some of their thoughts about the FSR&D approach. We have included in these readings papers by Richard Harwood, David Norman and Elon Gilbert, Donald Winkelmann and Edgardo Moscardi, Robert Hart, Hubert Zandstra, Peter Hildebrand, Jerry McIntosh, Bert Krantz, and Donald Plucknett.

These papers are but a sample of the contributors' writings on FSR&D. Nevertheless, they convey themes that run through many of their other writings. We of the FSR&D project team have benefited greatly from these and similar papers and perhaps even more so from direct personal contact with each member of this group.

For this help, we offer these writers our sincerest
thanks. We also wish to thank Jan Owen and Donald
Zimmerman for their editorial assistance; Hanae Akari for
the drawings; and Vicky Lynn, Christine Stanley, Margaret
Neff Withey, and Cheryl Buster for typing the manuscripts.

W. W. Shaner
P. F. Philipp
W. R. Schmehl
Fort Collins, Colorado

Introduction

W. W. Shaner

This book of readings contains some of the more recent thoughts by those actively concerned with farming systems research and development (FSR&D) methodology and its application. By FSR&D, we mean agricultural research and technology development that views the whole farm as a system and focuses on (1) the interdependencies among the components under the farm household's control and (2) how these components interact with the physical, biological, and socioeconomic factors not under the household's control.

The papers contained in this book of readings are by those practitioners who attended the FSR&D workshop in August 1979 sponsored by this project. Following, we first present brief summaries of these papers and then present the papers themselves. The papers, in order of their appearance, are by Richard Harwood, David Norman and Elon Gilbert, Donald Winkelmann and Edgardo Moscardi, Robert Hart, Hubert Zandstra, Peter Hildebrand, Jerry McIntosh, Bert Krantz, and Donald Plucknett.

The first paper, by Harwood, categorizes farming systems according to their stage of development and resource use. In doing this, Harwood uses conceptual layouts of farms based on the farmer's use of land and water resources and the farm's total productivity. From this categorization, plus close observation and measurement of farming activities, researchers can better understand farming enterprises. This understanding in turn aids researchers in identifying opportunities for improvements related to such topics as multiple cropping, home food production, and crop-animal interactions.

The second paper, by Norman and Gilbert, concentrates on conceptualizing farming systems research and then raises several methodological issues. The authors' categorization includes identifying technical and human elements both under the farmers' control and not under their control. Issues concern those such as "How holistic to make the analysis?" "Whose interests should be considered?" and "Which constraints should be taken as given?"

The third paper, by Winkelmann and Moscardi, describes some of the procedures developed and implemented by CIMMYT's[1] Economics Program. These procedures help in identifying farmers representative of particular environments and in designing technologies specifically to the farmers' needs. The approach centers on farming systems in which maize and wheat are important crops, seeks to develop effective collaboration between biological scientists and economists, and searches for relatively short-term improvements that are better than farmers' existing practices.

The fourth paper, by Robert Hart, describes a systems approach to the description and analysis of small farming systems. This approach draws on the integrative methodologies developed through study of ecosystems. Hart then develops a hierarchical framework starting with an agricultural region and ending with an individual crop or type of animal. He then traces the flows of money, materials, energy, and information into the system and the resulting outputs from the system.

In the fifth paper--a companion to the fourth paper--Hart illustrates his approach by using a small farm in Honduras. He reports on a year-long study of a farm family's activities and provides interesting insights into the way the family managed the farm.

The sixth paper, by Zandstra, contains the elements of IRRI's[2] approach to cropping systems research. Much of this work involves member countries of the Asian Cropping Systems Network. In his paper, Zandstra describes the interactions between the farmers' environment and management. This division into environmental and managerial factors has similarities to the Norman and Gilbert division of human and technical factors. Using his division, Zandstra then describes the essential steps in cropping systems research, which include site selection and description, cropping systems design and testing, and the application of results through pre-production testing.

The seventh paper, by Hildebrand, centers on his involvement with ICTA.[3] The ICTA approach has moved the focus of attention from the research station to the farmers' fields--where problems are identified through

---

[1] CIMMYT is the acronym for the Spanish wording for the International Maize and Wheat Improvement Center headquartered in El Batan, Mexico.

[2] IRRI is the acronym for the International Rice Research Institute in Los Banos, Philippines.

[3] ICTA is the acronym for the Spanish wording for the Agricultural Science and Technology Institute in Guatemala City, Guatemala.

reconnaissance surveys. Then, on-farm experiments are designed for farmers who follow similar farming practices. A key element of ICTA's approach is the emphasis on farmers' tests in which farmers control the experiment and evaluate the results. Another key element is ICTA's reliance on interdisciplinary teams of biological and social scientists.

The eighth paper, by Jerry McIntosh, describes the cropping systems research program in Indonesia. This country--part of the Asian Cropping Systems Network-- receives significant research help from IRRI. McIntosh's paper relates how Indonesia has used a cropping systems approach to help improve food production and to relocate farmers from crowded areas to unused, yet potentially productive lands. McIntosh also describes the approach to target and research area selection, research trials for alternative cropping patterns, and implementation of results.

The ninth paper, by Bert Krantz, describes ICRISAT's[4] general approach to farming systems work. This institution is exploring alternative agricultural systems for increasing and stabilizing agricultural production in the semi-arid tropics. ICRISAT's farming systems effort has concentrated on the problems of soil erosion, a limited and uncertain water supply, and the lack of suitable technology for these conditions. As a result, ICRISAT is developing technologies related to surface storage of water, erosion control, seedbed preparation, earth-shaping equipment, and related matters.

The last paper, by Donald Plucknett, recounts some of his experiences as a member of the CGIAR's[5] Technical Advisory Committee's review of farming systems research at the International Agricultural Research Centers. He stresses that learning about the farmer and the farmer's system and having a conceptual framework in mind leads to a better understanding of the reasons different organizations conduct research differently. Using this approach, the Committee categorized the Centers' efforts according to their relative emphasis on base data analysis, on-farm studies, and research station experimentation. Plucknett also stresses the importance of on-farm research, interdisciplinary teamwork, the search for practical solutions to farmers' problems, and the better use of available data.

---

[4] ICRISAT is the acronym for the International Crops Research Institute for the Semi-Arid Tropics headquartered in Hyderabad, India.

[5] CGIAR is the acronym for the Consultative Group on International Agricultural Research.

      Additional readings by these and other writers on FSR&D can be found in the lists of references in this book and in the references in the project's book, *Farming Systems Research and Development: Guidelines for Developing Countries.*

1
Farming Systems Development in a Resource-Limiting Environment

Richard R. Harwood

The Status of Third World Agricultural Development

"That the world food situation today is serious, even precarious, is well established" (Wortman, 1978). Increases in food production in the third world are, in good years, barely able to keep pace with rising demand. While tremendous advances have been made in the past 15 years through improved crop technology, the breadth of change has been disappointing. Recent estimates, for instance, indicate that 75 percent of the world's rice farmers, concentrated mostly in South and Southeast Asia, have not been affected by the new rice technology (Ponnamperuma, 1979). Others have decried the "disruptions in rural societies produced by almost exclusively production-oriented agricultural development of the past decade" (Anderson, 1979). Regardless of the viewpoint, the problems of third world agricultural development are today greater than ever.

Volumes have been written on the shortcomings of the Green Revolution. Ponnamperuma (1979) states that "Small farmers cannot provide the management inputs required to extract the high yield potential of modern varieties." We can summarize most of the rhetoric with the observation that resource limitations are responsible for much of the lack of progress. Shortages of cash inputs and mechanization or the money to buy them, lack of supporting infrastructure (roads, markets) and limited production potential (land, water, and favorable climate) vie with accusations of inappropriateness in new technologies for their share of the "blame." The idea that many of us had a few years ago of new varieties, proper inputs, and fair market prices being the main answer to farm production problems has been severely jolted if not completely invalidated. A recent summary of extensive nitrogen response studies across Asia (Ahsan, 1978) found that on the farms studied, net farm income was negatively related to the level of nitrogen fertilizer used by rice farmers in Pakistan and Sri Lanka, and not related at all to net

5

farm income in Bangladesh, the Philippines, and Thailand.
Isn't anything sacred these days?

Development Under Resource Constraints

A realistic look at the global energy situation tells
us that not only energy, but capital for development will
continue to be severely limited in the foreseeable future.
We can safely conclude that efficiency of resource use
will be the name of the game in agricultural development
in future years. The great majority of farmers will continue to have access to only a limited rural infrastructure.

Limited availability and high cost of production inputs will continue to reduce the impact of these inputs.
The need for resource-efficient technologies, often specific to well-defined production environments is the
challenge of today's development team. This applies,
according to some, to the agriculture of the developed
countries as well as to third world nations. *Muddling
Toward Frugality* by Warren Johnson is an excellent treatise
with this theme.

We thus come to the need for farming systems development strategies. Many of the resource-efficient technologies are concerned with the complementarity and integration of enterprises on a farm for effective use of scarce
farm resources. The knowledge of those interactions and
the ability to enhance their effects are the realm of
farming systems research. It implies a farmer-involved
approach. It implies an understanding of component
technologies and their interaction with gradients of the
physical, biological, and socio-economic environments of
a farming system. Those aspects of farming systems technology are the focus of our study here for these few days.

Development Stages of Third World Agriculture

Many generalizations are common in today's litera-
ture about "subsistence" farms, "small" farms, and
"modern" farms. These terms bear relation to the amount
of production resources available to a farmer as well as
to the degree with which he utilizes them. A breakdown
by farm development stage (Table 1) gives insight into
the conditions for technology acceptance on those farms
(Harwood, 1979).

Shifting cultivation is one of the most widespread
types of farming in the world. It is practiced on marginal land where sustained production of annual crops is
not possible without major nutrient input. Many forms of
shifting cultivation cover a broad spectrum of types,
with most types including a portion of the cropped land
in fixed agriculture. The fixed portion may include lowland rice, tree crop mixtures or sustained cropping on
small portions of land that may have a more productive

Table 1. Characteristics of development stages in agriculture (for farms with a relatively high level of resource use for their development stage).

	Shifting cultivation	Permanent agriculture (subsistence) less than 10% sales	10-50% sales	Commercial family farms over 50% sales	Corporate or state farms
Proportion of farmers involved		over 40%		less than 50%	less than 3%
Predominant labor activities					
Landclearing	x				
Tillage by hand	x	x	x		
Tillage by animal		x	x	x	
Tillage by machine				x	x
Animal tending		x	x	x	
Crop tending	x	x	x	x	x
Nutrient cycle		x	x		
Harvesting	x	x	x	x	x
Marketing			x	x	
Types of farming systems					
Monoculture crops	no	yes	yes	yes	yes
Intercropping	yes	yes	yes	rarely	no
Draft animals	none	yes	yes	yes	none*
Pigs untended	yes	no	no	no	none*
Poultry untended	yes	yes	yes	yes	none*
Complementarity of interactions between crops and between animals	slight†	very high	high	moderate	slight
Importance of farmstead to family nutrition	slight	very high	high	moderate	slight

* Animals and cultivated crops are usually not mixed on corporate farms in the tropics.

† Negative when animals compete with people for food.

soil. The remaining land may be of low productive potential because of steepness of terrain, lack of nutrients,
or an easily deteriorating soil type that is difficult to
manage even with good inputs.

For many of these areas, returns on high infrastructure development costs may be marginal because of the
limited physical production resource. Labor productivity
in the hand cultivation system with low value crops is
barely adequate for survival. Shifting cultivation areas
are not, however, in the forefront of development efforts
because of low visibility and lack of resources for the
costly infrastructure development which conventional
development models require. The vast technological distance between shifting cultivation and commercial agriculture is staggering. It is usually linked with strong
social custom.

As agriculture has evolved over the centuries, there
has been a gradual shift to a permanent and stable agriculture through development of nutrient cycling systems
made possible by the inclusion of animals in the system.
With cleared, permanent fields, animal draft power can be
used. Labor productivity, agricultural intensification,
and farm productivity increase markedly. In the isolated
hills of Nepal such systems reach extremely high levels
of intensification and productivity with absolutely no
market-derived production inputs. The system requires
off-farm grazing and forest land to provide a source of
nutrients for cycling into intensive production fields.
It is an extremely highly structured system when pushed
to its maximum, with intensive and crucial interactions
between farm system components. Its many elements, including intensive intercropping, nutrient cycling, diversified and highly developed mixed-planting homestead
areas, and a delicate crop-animal balance have evolved to
maximize productivity in an environment where external
resource use is being minimized (Moseman, 1976).

We consider our interest in permanent subsistence
systems essential for three reasons. First, they represent a vastly improved potential system for much of the
present shifting cultivation area, where infrastructure
development and direct transition to a commercial system
are in the distant future. Second, the fixed agricultural subsistence systems of certain areas such as Nepal,
are relatively numerous compared to those of other areas.
Elements of the former should fit the latter. Third, and
most important, elements of these subsistence systems
appear to fit beautifully into commercialized farming
systems where infrastructure is costly and not fully
available or where farm production potential is low and
off-farm income (which would push the system toward
commercialization) is minimal.

Much of the crop technology involving intercropping
may also be relevant to partially commercialized, re-
source-limiting situations (Harwood, 1976). The People's

Republic of China has very dogmatically followed a policy
of using such technologies in their development of agriculture under severe production resource limitations
(Plucknett, 1980). As industrialization progresses, these
technologies will have served their purpose as stepping
stones and will probably be gradually replaced by external
production inputs.

The commercialized farm, part of a rural sector which
is closely linked with industrialized portions of society
to the benefit of both, has, in the past, been our development ideal. Labor productivity in this model becomes
our final goal in improving rural well-being, and many of
the more labor-intensive elements of farm structure (especially the intensive enterprise interactions) are replaced
by less labor-intensive methods. This is highly evident
in Taiwan. Farm systems become less diversified and
greater dependence is placed on capital inputs. I question the relevance of this model for all development
situations.

Resource Use of Three Development Stages

I have presented conceptual models of three farms
representing two extremes and a middle-ground of development stages (Figures 1-3). The partially commercialized
farm represents the highest level of resource use with
limited land and scarce external support. This farm
represents a broad spectrum of farm types that have been
only marginally touched by modern development. The
essential elements of such a production system include a
nearly complete provision for family dietary needs
through self-sufficiency food crops and chickens for meat.
These family-oriented food crops may be grown on a portion
of the cash crop acreage which is devoted to staple
grains, as well as on a homestead or mixed-planting area.
The cash value assigned to this production is high, as it
substitutes for retail cash expenditures on a high-margin
market. Its cash value relative to land resources used
is also high because of the high nutrient status of the
homestead area where the crops are grown. The importance
of this self-sufficiency production has been stressed by
Harwood (1979) and Martin (1978). Anderson (1979) has
done the most complete description of Asian systems.

Cash crop production has been the focus of most
development research. Cash crops are an important part
of the semi-subsistence farm, but their potential for
increase is often limited without the provision of greater inputs or better markets. Crops grown solely for
animal feed are rare, but where the market for meat is
good, some grain may be fed.

Animals include the free-ranging chickens which are
used mostly for home consumption. Their cash value is
likewise high relative to resources used because of their
limited competition with crops. These chickens are

Fig. 1.   Resource use: Productivity of a land-limited,
          partially commercialized farm.*

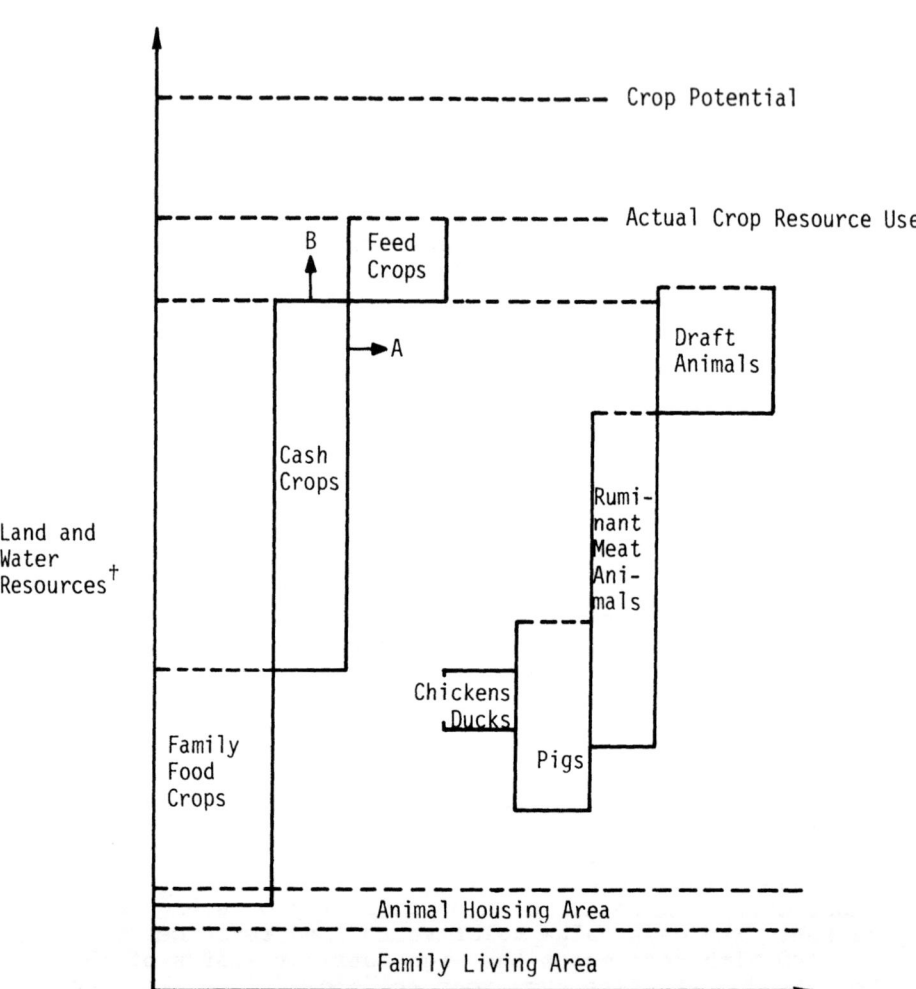

Total Annual Productivity (Cash Value)

*Value of an enterprise is indicated by its width, land resource use
by height. Enterprises which overlap on a horizontal line share
land/water resources.

†Subject to temperature and other limitations.

Note: A represents higher value of cash crops, e.g., vegetables if the
       market is available, and B represents greater land and water use.

Fig. 2. Resource use: Productivity of a shifting cultivation farm.

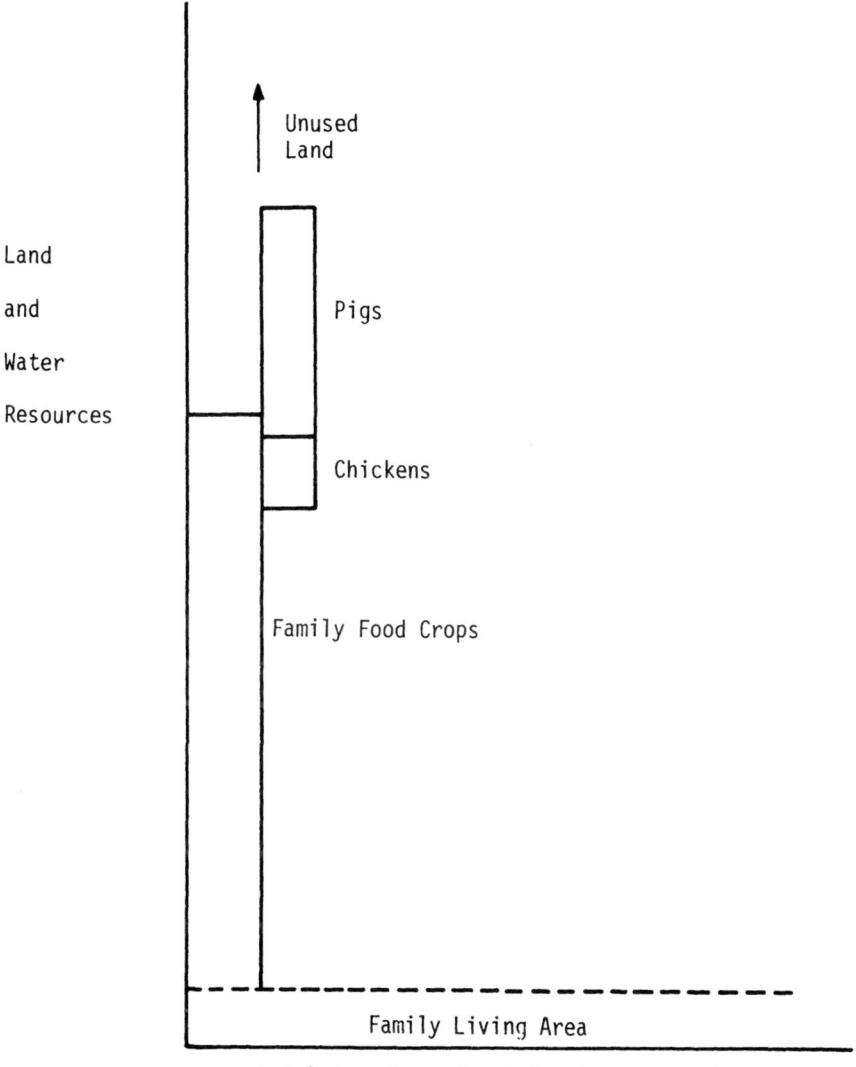

12

Fig. 3. Resource use: Productivity of a fully commercialized farm.

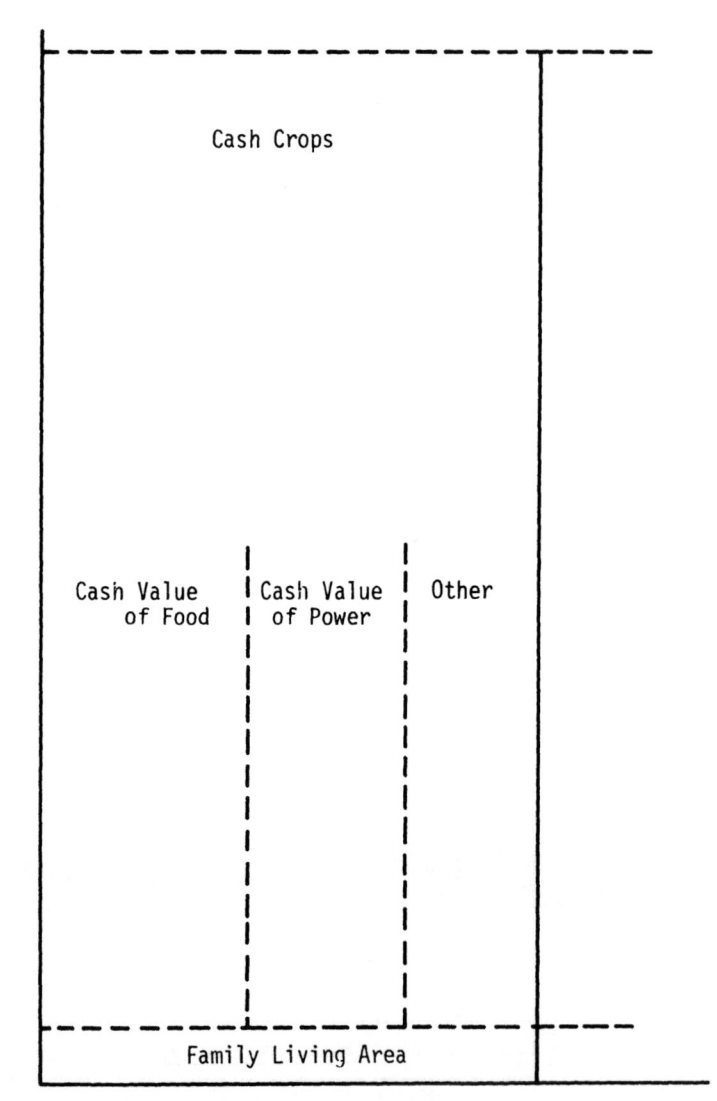

scavengers and utilize little, if any, of the farm's production resource. Pigs are less frequently found, but are always under confinement on such a farm. Their value is relatively high. They compete to some extent with ruminant animals for feed, but utilize mostly crop residues. They are an extremely important part of Chinese vegetable farming systems. Ruminants, both for meat and for draft are likewise complementary to the crop enterprises. In small numbers, they utilize weeds or crop residues. Their numbers are limited by the amount of feed available. Draft animals on the small farm have a higher value relative to the resources used primarily because of the replacement cost of their power, either in terms of human labor or of mechanical power. On a well-integrated farm, the animals thus utilize little of the marketable crop produce. They increase the farm productivity greatly with little additional input requirement. All of this assumes, now, that land and not labor is the limiting resource.

A second farm type, representing an early development stage, is the shifting cultivation farm. This farm is relatively unstructured and is more land-extensive. It has few complementary crop-animal relationships. Its crop productivity is low in relation to land resources used.

The third type, that of the fully commercialized farm, represents a labor-limiting situation with full access to inputs and markets. The system has few complementary interactions, lacking especially the crop-animal interactions which are labor-intensive. Total productivity per unit of land resource is lower because of the lack of these complementary enterprises. If land becomes a limiting factor, this type of operation will continue to give a high return on labor, but the farm family may have less net income. A large portion of the production will be used to purchase food or mechanical power, with its attendant high service costs in a developing economy. Many of the beneficiaries of our modern technologies have been those farmers with access to the resources to support such a farming system. Unfortunately, our efforts have been largely restricted to this high cash flow, "commercialized" philosophy stemming from our commodity rather than system development orientation.

The Need to Target According to Development Stage

It should be obvious that technology requirements are not only specific to physical environment but to development stage. It should also be obvious that certain types of farming systems may be more suited to different resource combinations. By orienting our farming systems research as well as our improvement efforts to specific development stages, we can begin to improve our understanding and our effectiveness in dealing with farm problems. This, then, is my first conclusion.

The Descriptive Process in Farming Systems Research

Once a target area has been selected, the systems
survey and description begins. This is not only the most
crucial phase but the one least successfully accomplished
in most systems research. Unfortunately, the systems are
so complex and so variable that precise enumeration is
extremely difficult if not impossible. Our penchant for
"hard"data and accuracy leads us to begin feverishly to
measure rather than to observe. We usually end up with
exhaustive detail about parts of the system, but we never
can put the whole picture together. We should always
start with the conceptual layout of a farm, perhaps not
unlike those of Figures 1-3. For each farm we could then
sketch in rough numbers for the major components, quantifying as much as possible the interactions. Above all,
this should be done quickly. The entire process for a
given target area should take no longer than a few days.
The timing of the *sondeo* method used by Hildebrand is outstanding. This descriptive phase should convey a clear,
if partially conceptual and subjective impression of the
entire farming system. Specific aspects of the system
can then be described in greater detail. The need for a
conceptual overview of the farming system types in the
target area as the first step in the descriptive process,
then, is the second conclusion.

Areas for Greatest Gains in Resourse Use Efficiency

There has been considerable attention given in
recent years to multiple cropping research. With new
short-season varieties, better weed control methods, and
improved methods for efficient water use, the potential
for increased cropping intensity has grown. A reawakened
interest in traditional relay and intercrop methods will
lead to further increases.
A second area is that of home food production. This
area has been nearly completely overlooked in recent
years. A third area is that of effective crop-animal
interactions for feed, power, and nutrient cycling.
Small animal production for home consumption can make a
substantial contribution to limited resource productivity.
A third conclusion, then, is the need to look at technologies not usually dealt with in development work.

Limitations

In order to realize these opportunities, several
problems may need attention--including the need for
security to prevent cattle rustling--animal confinement
laws to permit structuring of compatible crop-animal
interactions, markets for animal products, low-input

insect and disease resistant varieties, new crops for
multiple-crop sequences, seeds or planting materials for
homestead gardens, and many more. Finally, there is a
real need for a more thorough understanding of the fit of
crop and animal technologies to environmental gradients.

Conclusions

The emerging farmer-participant farming systems
methodologies are, for the first time, permitting us to
diagnose the more complex farm development problems and
to accurately target technologies to meet those needs.
We are beginning to institutionalize the heretofore mainly artistic skills of the highly successful development
scientists of the past. We must not become confused or
discouraged by the complexity of our undertaking, but
frequently stand back to assess our progress and regain
our bearings as we venture onto uncharted ground.

References

Ahsan, E. and V. Shotelersuk. 1978. Socio-economic
    factors relating to fertilizer use in rice farms of
    five Asian countries--Bangladesh, Pakistan, Philip-
    pines, Sri Lanka, and Thailand: a review of INPUTS
    trial IV, *In* Proceedings: Second Review Meeting, IN-
    PUTS Project, May 8-19, 1978. East-West Center,
    Honolulu, Hawaii. pp. 291-305.
Anderson, J. N. 1979. Traditional home gardens in
    Southeast Asia: a prolegomenon for second generation
    research. International Symposium of Tropical
    Ecology. Kuala-Lumpur, Malaysia.
Harwood, R. R. 1978. Agronomic and economic considera-
    tions for technology acceptance. American Society
    of Agronomy, Madison, Wisc.
Harwood, R. R. 1978. The application of science and
    technology to long-range solutions: multiple crop-
    ping potentials. *In* nutritional and agricultural
    development, Strimshaw, N. S. and M. Behard, (eds.).
    Plenum, New York. pp. 423-441.
Harwood, R. R. 1979. Small farm development: under-
    standing and improving farming systems in the humid
    tropics. Westview Press, Boulder, Colo. 154p.
Hildebrand, P. E. 1979. Summary of the *sondeo* methodol-
    ogy used by ICTA. Instituto de Ciencia y Tecnología
    Agrícolas. Guatemala.
Johnson, W. 1978. Muddling toward frugality. Sierra
    Club Books, San Francisco, Calif. 252p.
Martin, F. W. 1978. Planning the small subsistence farm
    for complete nutritional independence (mimeograph).
    Mayagüez Institute of Tropical Agriculture,
    Mayagüez, Puerto Rico.
Moseman, A. J., (ed.). 1976. A study of hill agricul-

ture in Nepal. The Rockefeller Foundation, New York. 124p.

Plucknett, D. L. et al. 1981. Vegetable farming systems in The People's Republic of China. D. L. Plucknett and H. L. Beemer, Jr. (eds.). Westview Press, Boulder, Colo. Published in cooperation with the National Academy of Sciences.

Ponnamperuma, F. N. 1979. IR42: a rice variety for small farmers of South and Southeast Asia. Agronomy Abstracts: American Society of Agronomy, Madison, Wisc. p. 46.

Wharton, C. R., Jr. 1969. Subsistence agriculture: concepts and scope. *In* subsistence agriculture and economic development, C. R. Wharton, Jr., (ed.). pp. 12-20. Aldine, Chicago.

Wortman, S. and R. W. Cummings, Jr. 1978. To feed this world. Johns Hopkins. p. 1.

2
A General Overview of Farming Systems Research

D. W. Norman
Elon Gilbert

The first part of this paper presents, in summary form, definitions of a farming system (FS) and farming systems research (FSR) and includes a brief review of the types of FSR currently in existence. The second part of the paper is devoted to the methodological and implementation issues associated with deriving immediate solutions to farmers' problems.

The Farming Family (Household) and Its Environment

In most types of agriculture in less developed countries (LDCs) the unit of production (the FS) and the unit of consumption (farming household) are intimately linked and cannot be separated. The specific FS adopted by a given farming household results from its members, with their managerial know-how, allocating the three factors of production (land, labor, and capital) to three processes (crops, livestock, and off-farm enterprises) in a manner which, with the knowledge they possess, will maximize the attainment of their goal(s).

The FS is determined by the environment in which the farming family operates. The "total" environment in which it operates can be divided into the technical (natural) and human elements (see Fig. 1).

The technical element reflects what the potential farming system can be and therefore provides the necessary condition for its presence. The technical element can be divided into: physical factors (water, soil, solar

---

1 No attempt has been made to cite specific references in the paper. A selected list of references--by no means complete--is given at the end of the paper. In addition, the paper benefits greatly from many other references and comments from 24 reviewers of the first draft of a recent review of FSR (Gilbert et al., 1980).

18

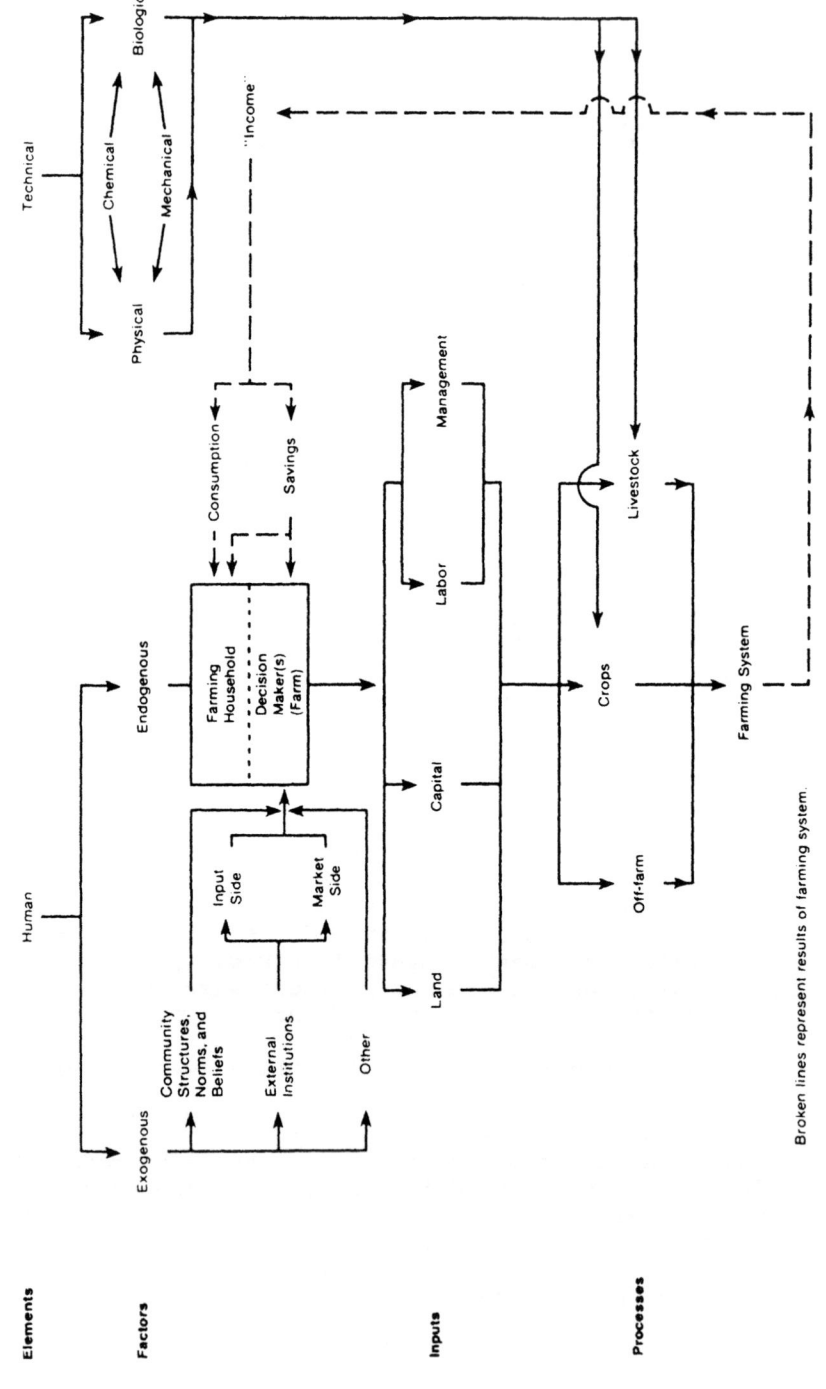

Fig. 1. Schematic representation of some determinants of the farming system.

radiation, temperature, etc.) and biological factors
(crop and animal physiology, disease, insect attack, etc.).
Technical scientists have been able to modify the technical elements to some extent.

The human element has often been neglected in traditional research approaches to developing improved technologies. This accounts for the technologies often being
rejected or at best being differentially adopted, thereby
resulting in an inequitable distribution of benefits.
The human element provides the sufficient condition for
the presence of an FS which is a subset of the potential
productive activities defined by the technical element.
The interaction of the technical and human elements
determine what the actual farming system will be.

The human element can be divided into two components
or groups of factors. The exogenous factors--the social
milieu in which the farming household operates--are
largely out of the control of the individual farming
household but will influence what its members are able to
do. These factors can be divided into three broad groups:
(1) community structures, norms, and beliefs, (2) external institutions, which include those influencing farming
decisions related to supplies of inputs and markets for
the farmers' commodities, and (3) other factors, such as
farm location and population density. On the other hand,
endogeneous factors (land, labor, capital, and management)
are under control of the individual farming households
and can be used by them to derive an FS consistent with
their goal(s) subject to the boundary conditions laid
down by the technical element and exogenous factors. The
endogenous factors can, under certain circumstances, be
complemented and supplemented in quantitative and qualitative terms through the influence of exogenous factors
such as capital through a credit program and management
via extension.

Objective of the FSR Approach

The primary objective of FSR is to improve the well-
being of individual farm families by increasing the overall productivity of the FS in the context of the entire
range of private and societal goals and given the constraints and potentials imposed by the technical and
human elements which determine the existing farming systems.

Increased productivity is achieved through two types
of developmental strategies. The first is the development and dissemination of relevant improved practices
(technologies). The second involves changes in the exogenous factors either to create opportunities for certain
types of improved production systems to be adopted by
individual farming families, or to provide conditions
conducive to the adoption of technologies already available. Examples are: encouragement of group activities on

the part of farmers (to enable watershed management to be effective); and influencing necessary adjustments in agricultural policies and actions of farmer contact agencies.

To date, work in FSR has been largely confined to developing improved crop technologies. The second type of strategy has not as yet been generally linked to FSR. Therefore, this potential role of FSR still has to be demonstrated to be of practical value (due in part to resistance to the "bottom up" characteristic of FSR).

Defining and Operationalizing FSR

Whether or not it is explicitly called FSR, research can be considered farming systems research if it has the characteristics discussed below.

First, the farm as a whole is viewed in a comprehensive manner with a recognition of the interdependencies and the interrelationships within the natural and human environment in which the farming system is operated. As such, it is more holistic in orientation than the reductionist approach traditionally used by technical agricultural scientists. The latter approach has involved studying one or two factors at a time while attempting to control all others. The inclusion of the perspective of the whole farm in the research process means that explicit attention is focused on such characteristics as goals, components, and constraints of the farming systems that are present.

Second, the choice of priorities for research reflects the initial study of the whole farm.

Third, the farming system can be broken down into a number of subsystems which may overlap and interact with one another. It is legimate to consider research on a subsystem as being FSR provided the connections with other subsystems are recognized and taken into account.

Fourth, the evaluation of the results and their implementation take the linkages between the subsystems explicitly into account.

The methodological complexities of undertaking FSR can be great because of its systems focus and its "holistic" characteristic. Therefore, in practice, in order to make it operational, advantage is taken of the characteristics of FSR mentioned above. In other words, the concept of the "total" environment is preserved, but instead of assuming that all factors determining the actual farming system can be potential variables, subject to manipulation, some are treated as parameters. In addition to methodological considerations, the mixture of variables and parameters is influenced by such factors as the mandates of the institutions involved, the effectiveness of linkages with other institutions, and the resources available (i.e., time, skill, and finances). FSR may be called FSR in the small if the number of variables is small relative to the number of parameters; or FSR in the large if the number of variables is large with relatively few parameters.

Types of FSR Programs

As well as FSR programs being differentiated on the basis of the ratio of variables to parameters, they can be classified in the following ways.

First, "upstream" types of FSR programs have a developmental orientation and usually do not provide results for immediate adoption by farming families. Perhaps more aptly called resource management research, "upstream" FSR programs involve using a systems approach on experiment stations to provide prototype solutions aimed at alleviating major constraints to agricultural improvement. Examples include the watershed management research by ICRISAT and the research on minimum tillage at IITA. Along with the results from commodity improvement programs, they contribute to the body of knowledge (Fig. 2) and are available for feeding into the "downstream" types of FSR programs.

Second, "downstream" FSR programs, which are the main concern of this paper, have an adaptive orientation and aim at developing and introducing strategies that will improve the productivity of farming systems for target groups of farming families in the short run. This requires selectively drawing upon available information (i.e., body of knowledge in Fig. 2) in the process of designing practices or recommendations for a particular farming system on the basis of an analysis of the constraints of that system. Therefore, recommendations are produced which are suited to a specific local situation. This involves working directly with farmers (i.e., on-farm research) and as a result, reducing to a minimum work on the experiment station.

FSR type programs are now expanding rapidly throughout the world and are being undertaken at national, regional, and international institutes. Both types of FSR programs mentioned above are important. The relative degree of emphasis on one or the other will depend upon various considerations including the nature of the problem and the research resources available. "Upstream" type FSR programs are necessary when traditional reductionist research approaches cannot contribute to solving the problem, which leaves a gap in the body of knowledge and inhibits the ability of "downstream" FSR to produce appropriate or relevant practical strategies for farming families in the short run. However, the research resources required to undertake "upstream" FSR programs are often great, and generally result in such activities being concentrated in regional or international institutes. On the other hand, "downstream" FSR programs, with their focus firmly on the needs of specific groups of farming

Fig. 2. Schematic framework for farming systems research at the farm level: Downstream farming systems research.

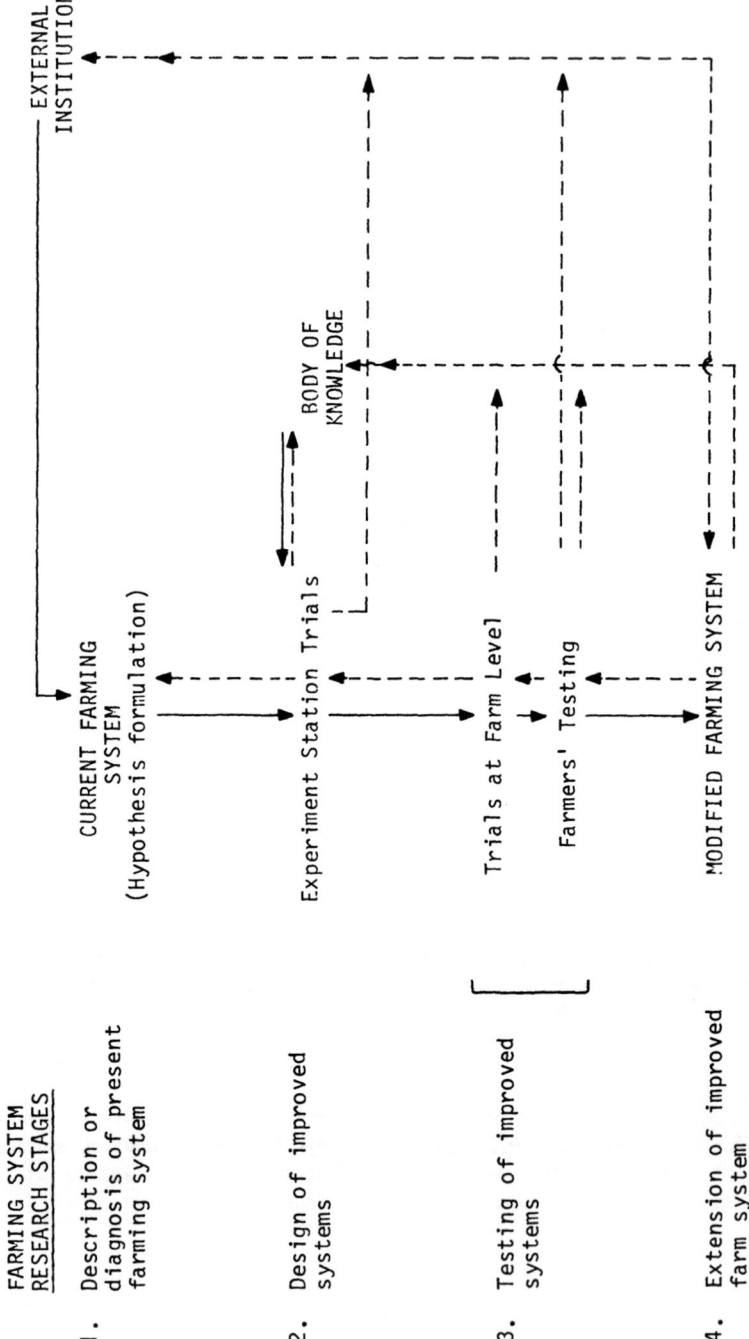

FARMING SYSTEM
RESEARCH STAGES

1. Description or diagnosis of present farming system

CURRENT FARMING SYSTEM (Hypothesis formulation)

2. Design of improved systems

Experiment Station Trials

3. Testing of improved systems

Trials at Farm Level

Farmers' Testing

4. Extension of improved farm system

MODIFIED FARMING SYSTEM

EXTERNAL INSTITUTIONS

BODY OF KNOWLEDGE

families, have a comparative advantage in being located
in national institutions. Therefore, the effectiveness
of "upstream" FSR depends to an important extent on the
strength of linkages with one or more "downstream" programs operating in specific locations. "Downstream" FSR
can be a useful event where the body of knowledge is not
well developed since it can assist in defining research
priorities for "upstream" programs as well as commodity
and discipline oriented programs thereby improving the
likelihood that these programs will produce relevant
research results.

Stages and Attributes of "Downstream" FSR

     A conventional wisdom is emerging about "downstream"
FSR, although there are many differences in details of
methodology and implementation. Some of the attributes
of "downstream" FSR include the following:
     First, there are four successive stages in the
research process (description, design, testing, and extension). The descriptive or diagnostic stage is undertaken to determine constraints, needs, and flexibility in
the current farming system. This provides an input into
designing, testing, and extending improved strategies,
whose potential suitability will be determined by the
application of appropriate evaluation criteria ascertained during the descriptive stage.
     Second, the objectives of the farming family are
directly incorporated into the research process. The
farming family is the central unit in the research process
being directly involved in the descriptive, testing, and
extension stages. Testing consists of trials at the farm
level (i.e., under the direction of the research team
with the farmer participating) and farmer tests (i.e.,
totally under farmer control). Involvement of farmers
gives them a "voice" in the research process and ensures
the use of evaluation criteria relevant to them. Evaluation criteria for the adoption of improved practices for
the farming family relate to the family's ability to
adopt a specific practice (necessary conditions) and its
willingness to do so (sufficient conditions). The necessary conditions include technical feasibility, social
acceptability, and compatibility with external institutions and support systems. Obviously, the necessary conditions will influence the sufficient conditions, namely
the willingness of the farming family to adopt a specific
practice. Sufficient conditions include compatibility of
the practice with the goal(s), such as self-sufficiency
in staple foods and profit maximization of the farming
family and the farming system currently practiced.
     Third, efforts are made to incorporate community and
society needs into the FSR process by trying to ensure a
convergency between private (often short run) and societal (usually longer run) interests. Examples of possible

conflicts would be where satisfying short run needs of individual farming families would result in long run societal costs such as degradation of the natural resource base and increased inequalities in welfare distribution. It is necessary to develop improved strategies that will avoid such conflicts.

Fourth, the FSR approach, by including farmers, taps the pool of knowledge in the society and enables research and hence developmental strategies to build upon the good points of the present farming systems, while at the same time minimizing the time spent in "rediscovering the wheel" (e.g., the value of intercropping).

Fifth, FSR recognizes the locational specificity of the technical and human (exogenous and endogenous factors) elements. This requires disaggregating farming families into homogeneous subgroups (recommendation domains) and developing strategies appropriate to each. Farming families in a particular subgroup will tend to have similar farming activities and to include similar social customs, similar access to support systems, comparable marketing opportunities, and similar present technology and resource endowment.

Sixth, the whole farm perspective of FSR compels the adoption of an integrative function which increases the potential for exploiting complementary and supplementary relationships between resources and enterprises, and the derivation of solutions compatible with the needs and capacities of farming families. The systems farmers traditionally practiced recognize such relationships (e.g., crops and livestock, staggered planting dates, etc.). To ensure that the integrative and beneficial relationships are being adequately considered and exploited, requires a multidisciplinary team (both technical and social scientists) working together at all four stages of the research process.

Seventh, the process of FSR is recognized as being dynamic and iterative with links in both directions between farmers, research workers, and funding agencies rather than simply the presence of forward links characteristic of the "top-down" approach. The iterative characteristic can improve the efficiency of the research process by providing a means to fine-tune improved technologies for a specific locale.

Eighth, FSR, unlike reductionist research approaches, has a wider perspective and is concerned with the productivity of the entire farming system. Therefore, rather than just being concerned with technical issues it can also encompass nontechnical or institutional issues through influencing the exogenous factors, as has been done in the Caqueza project in Colombia and the Technological Package project at Central Luzon State University in the Philippines. The latter project is addressing not only issues with respect to increasing production but also the related issues of marketing and processing.

Finally, FSR complements and does not compete with other research approaches. For example, reductionist commodity based research programs provide essential inputs into the body of knowledge (Fig. 2) which "downstream" FSR relies on for facilitating quick results at specific locations. Also, as mentioned above, the application of "downstream" FSR can help redefine or refine research priorities in other types of research programs.

The Role of Social Scientists in "Downstream" FSR

The preceding section argued that a multidisciplinary team consisting of technical and social scientists is required to undertake "downstream" FSR. To be effective, such teams need to work in an interdisciplinary manner, that is, different disciplines working together rather than independently on a specific problem. The interdisciplinary approach assists in understanding the relationship between the technical and human elements; for example, whether late planting of a crop is due to climatic conditions, lack of available labor, or a risk aversion strategy against losses from early planting. An understanding of the reason(s) is important as an input into designing and testing relevant improved developmental strategies.

The role of social scientists in "downstream" FSR will vary according to the stage of the research process and the stage of development of the target groups of farming households. Improved developmental strategies should be compatible with the goal(s) of the farming family. However, the objective function of farmers and therefore what motivates them will change as they move from a subsistence type of farming system to one that is more commercialized. In the case of subsistence FS, understanding the goal(s) may be a particularly complex task while in the case of the commercialized FS, the goals may be easier to articulate (profit maximization). For farming families who are near the self-sufficient end of the spectrum, resources (social scientists) will need to be devoted to understanding just what the goal(s) are, while the closer the farming families are to practicing a fully commercialized system of agriculture, the more emphasis is likely to be placed by the social scientists on work connected with the external institutional support system.

Some Methodological Issues of "Downstream" FSR

Due to the fact that the methodology for undertaking "downstream" FSR is still going through a period of evolution, a large variety of methodological issues requires resolution. Not surprisingly, perhaps, there are often considerable differences in opinion as to how severe they are and how they should be dealt with. Some of the most frequently mentioned methodological issues are as follows.

First, how holistic should FSR be? As mentioned earlier the methodological problems increase as the FSR program becomes more holistic (i.e., the ratio of variables to parameters becomes higher). In addition, the present state of the art of undertaking FSR means that most current work is on the crop process and is largely confined to development of improved technologies. Practical problems also restricting the scope of "downstream" FSR are the mandates of institutions in which they are located (i.e., usually technical crop research institutes) and poor or weak links with other research institutions and with policy-making and farmer contact agencies. Related to the question of how holistic "downstream" FSR should be is the issue of whether the policy-institutional environmental factors should be treated as parameters or variables. Increasingly, it is being suggested that these factors might be treated as variables subject to manipulation, as suggested earlier. This micro-macro link is important in maintaining the viability of "downstream" FSR in the long run through the added dimension it gives to creating conditions conducive to improving the productivity of farming systems and hopefully the welfare of farming families.

Second, what needs or constraints are to receive focus in the research process? Should they be those articulated by farming families (i.e., felt needs), those scientifically ascertained by research workers, or those reflecting the needs of society? As discussed earlier, criteria used in developing improved strategies should reflect the felt needs of farming families providing they are not incompatible with the needs of society (e.g., there is not a decline in soil fertility, nutritional levels, increasingly inequitable income distribution, etc.). Strategies developed need to ensure convergence between short run private interests and those of the society in the long run. Although there is, in principle, agreement with the above, there is often disagreement as to how societal interests can be incorporated practically into "downstream" FSR. The problem of doing this relates to the methodological complexity of their incorporation and the time that would be required in deriving societal impact evaluations.

Third, the needs or constraints that are identified may be technical, economic, or socio-cultural in nature. What approach should be used in dealing with them? Two approaches are generally used. The first is accepting the constraint and developing strategies that exploit the flexibility that exists in the current farming system while at the same time not further exacerbating the constraint. Socio-cultural constraints should not generally be broken. The second is developing strategies that will

overcome the constraint. The decision about which approach to use generally depends upon the constraint sever-
ity, flexibility that exists in the current farming system, availability of potentially improved strategies
either to break the constraint or to exploit the flexibility, compatibility with societal goals, etc.

Fourth, is it necessary for "downstream" FSR to be
expensive? It is viewed by some to be expensive because
of its locational specificity and thus the need to focus
on limited numbers of farmers. The expensive nature is
emphasized because of the opportunity costs of neglecting
other farmers. Therefore, the quest for minimizing costs
in the research process is a major issue. Considerable
controversy exists concerning the degree to which costs
can and should be reduced and the ways in which this
should be done. In general, three approaches are being
used to try to minimize costs. Seeking ways to reduce
time and resources required for moving through the four
research stages is the first. Methods used should be
based on the criterion of the lowest possible cost com-
mensurate with the degree of understanding that is necessary. Can this be done with base data analysis plus an
informal exploratory (sondeo) survey and a one-shot formal
survey? Or is a detailed twice weekly formal survey required for a period of one year? Can modelling techniques help improve understanding, or does this come at
too high a cost? In the testing stage, should farmers be
selected that are the better farmers, the most cooperative farmers, or simply the representative farmers? Representative farmers may not, for example, be so cooperative which would reduce the efficiency and effectiveness
of dialogue and the timely conclusion of the testing
stage. Considerable controversy still exists concerning
the way in which these and other questions should be
resolved in the interests of minimizing costs and time.
Finding ways to maximize the return from the location-
specific nature of "downstream" FSR by determining the
transferability of the results to other similar "total"
environments is the second approach. Introducing some
flexibility into the improved practices increases the
potential of transferability but this may come at some
cost in terms of the potential level of return. Is this
desirable or not? Controversy exists with respect to
this. The last approach is seeking the best of readily
available solutions, that is, "better but not necessarily
best" or "non-perfectabilitarian." How much fine tuning
should there be thereby extending the length of the testing stage?

Fifth, in terms of developing improved practices
(technologies), should emphasis be placed on single trait
innovations, or should the complementary or synergistic
effects between the various components in packages of
improved practices be exploited? In theory, the latter
is desirable, but in practice the former is much more

common. A possible compromise is to design and develop packages of improved practices that permit, in an explicit manner, a stepwise approach to the adoption of the various components of the package.

Some Implementation Problems of "Downstream" FSR

Credibility problems in terms of both practical results (i.e., incremental and not spectacular although hopefully pervasive) and professional respect (i.e., by peers of own discipline) can result in difficulty of attracting adequate resources for FSR.

Intra-institutional adjustments to accommodate "downstream" FSR programs also can be difficult. Traditionally research programs have been organized along discipline lines and more recently on the basis of commodities. FSR means crossing both discipline and commodity lines. Narrow mandates and poor links cause problems and sometimes necessitate work on only one process or even part of that process. Cooperation between technical and social scientists may be difficult if they are not working within one institution--which unfortunately is often the case.

Another consideration is that links between FSR programs in regional, national, and international research institutions need rationalization to exploit the advantages of each. National programs have advantages in emphasizing downstream FSR although the problems mentioned above can be difficult to resolve in practice at this level. Also, the links in national programs with agricultural policy and farmer-contact agencies are generally weak, therefore making it difficult for FSR to play a constructive role in rural development programs. Links of FSR programs in regional and international institutes through networks are important in providing justification for their "downstream" FSR programs and outlets for FSR programs of an "upstream" type for which they have a comparative advantage.

Finally, there is the problem of identifying suitable individuals to participate in FSR programs. Training programs in FSR currently available are short term in nature and offered at international and occasionally at regional and national institutions. FSR training in formal degree programs is not available. In theory, a developed country's institutions might assist, but at present few staff have firsthand experience in FSR. Further field experience in FSR should be an important part of the training program which is not easily obtained in developed country institutions. Linkages between developed country institutions and FSR programs in LDCs could be important in facilitating practical experience for students and faculty alike.

References

Asian Cropping Systems Working Group 1979. Network methodology and cropping system research in Indonesia. Central Institute for Agriculture, Bogor, Indonesia.

Binswanger, H. P. and J. G. Ryan, 1979. Village level studies as a focus for research and technology adaptation. (Paper presented at the International Symposium on Development and Transfer of Technology for Rainfed Agriculture and the SAT Farmer, Hyderabad, India, Aug. 1979.) ICRISAT, Hyderabad, India.

Byerlee, D., S. Biggs, M. Collinson, L. Harrington, J. C. Martinez, E. Moscardi, and D. Winkelmann, 1979. On-farm research to develop technologies appropriate to farmers. (Paper presented at the Conference of the International Association of Agricultural Economists, Banff, Canada, Sept. 1979.)

Byerlee, D., M. Collinson, R. Perrin, D. Winkelmann, S. Biggs, E. Moscardi, J. C. Martinez, L. Harrington, and A. Benjamin. 1980. Planning technologies appropriate to farmers: concepts and procedures. El Batan, CIMMYT, Mexico.

Collinson, M. P., 1979. Micro-level accomplishments and challenges for the less-developed world. (Paper presented at the Conference of the International Association of Agricultural Economists, Banff, Canada, Sept. 1979.)

Elliott, H., 1977. Farming systems research in Francophone Africa: methods and results. (Paper presented at the Middle East and Africa Agricultural Seminar of the Ford Foundation, Tunis, Tunisia, Feb. 1-3, 1977.)

Gilbert, E. H., D. W. Norman, and F. Winch, 1980. Farming systems research: a critical appraisal. MSU Rural Development Paper No. 6, Michigan State University, East Lansing, Michigan, 1980.

Hildebrand, P. E., 1979. Generating technology for traditional farmers--the Guatemalan experience. (Paper presented at the International Congress of Plant Protection, Washington, D. C., August 5-11, 1979.)

IRRI (ed.). 1977. Cropping systems research and development for the Asian rice farmer. IRRI, Los Banos, Philippines.

ISRA. 1977. Recherche et développement agricole: les unités expérimentales du Sénégal. ISRA, Dakar, Senegal.

Jodha, N. S., M. Asokan, and J. G. Ryan, 1979. Village study methodology and resource endowments of the selected villages in ICRISAT's village level studies. Occasional Paper No. 16. ICRISAT, Hyderabad, India.

Kampan, J., 1979. Farming systems research and technology for the semi-arid tropics. (Paper presented at the International Symposium on Development and Transfer of Technology for Rainfed Agriculture and the SAT

Farmer, ICRISAT, Hyderabad, India, August 1979.)

Menz, K. M., 1979. Unit farms and farming systems research: the IITA experience. Discussion Paper No. 3/79. IITA, Ibadan, Nigeria.

Navarro, L. A., 1979. CATIE's development orientated agriculture research effort on the Central American Isthmus. (Seminar given at the University of British Columbia, Vancouver, Canada, Sept. 1979.)

Norman, D. W., 1979. The farming systems approach: relevancy for small farmers (In Karaspan, A. S. (ed.), Increasing the productivity of small farms. CENTO, Lahore, Pakistan.) p. 133-152.

Okigbo, B. N., 1979. Cropping systems in the humid tropics of West Africa and their improvement. (Paper presented at the IITA Research Review, IITA, Ibadan, Nigeria, 1979.)

Ryan, J. G. and H. P. Binswanger, 1979. Socio-economic constraints in the semi-arid tropics and ICRISAT's approach. (Paper presented at the International Symposium on Development and Transfer for Rainfed Agriculture and the SAT Farmer, ICRISAT, Hyderabad, India, August 1979.)

Technical Advisory Committee, 1978. Farming systems research at the international agricultural research centers. Technical Advisory Committee, Consultative Group on International Agricultural Research, Washington, D. C.

Winkelmann, D. and E. Moscardi, 1979. Aiming agricultural research at the needs of farmers. (Paper presented at the Seminar on Socio-Economic Aspects of Agricultural Research in Developing Countries, Santiago, Chile, May 7-11, 1979.)

Zandstra, H. G., 1979. Cropping systems research for the Asian rice farmer. Agricultural Systems, 4: 135-153.

3
Aiming Agricultural Research at the Needs of Farmers

Donald Winkelmann
Edgardo Moscardi

The Problem

Few farmers in developing countries are following the recommendations of researchers and extension workers. Explanations for this difference between practice and recommendations abound.

Some claim that farmers are at fault, arguing that preferences based on traditionalism lead farmers to reject unfamiliar technologies. Some point to extension, arguing that the utility of improved technologies has not been demonstrated to farmers. Others claim that inadequate credit limits farmers' ability to adopt technologies. Some emphasize that imputs are not available in a timely way and at appropriate prices. Finally, but less frequently encountered, some contend that recommended technologies are often not appropriate for farmers.

Certainly each of these explanations has been valid for some time and place. However, a number of recent experiences has shown even the poorest farmers--presumably among the most tradition-bound and usually among those with least access to inputs, information, and markets--taking up certain technologies while rejecting others. These experiences suggest that more attention should be given to the adequacy of recommended technologies. This, in turn, implies that more attention be given to the research systems which develop technologies.

In 1974 the International Maize and Wheat Improvement Center's (CIMMYT) Economics Program initiated its work to identify effective procedures for developing technologies. That effort involved collaboration with professionals in national programs and with CIMMYT staff assigned to regional and to national programs. At headquarters, economics joined with the maize and wheat training programs in pursuing work in procedures. The following discussion is based on our interpretation of those experiences.

The procedures which have emerged are now being tried in several national maize and wheat programs. They emphasize identifying the production problems of representative farmers and integrating the critical dimensions of their decision making into research on new technologies.

This concentration on research does not imply that the other issues mentioned earlier are not important. They are. The intention here is to add emphasis to the importance of the research system, to its procedures and its product.

CIMMYT's interest in such procedures relates directly to the Center's association with national programs. The Center is a producer of intermediate goods--elements of new technology, training, and procedures--which national programs apply in forging improved technologies. The procedures in this case relate precisely to the process from which improved technologies emerge.

Characteristics of Useful Technology

The utility of technologies can be judged from two related perspectives: that of the farmers and that of the larger society. In most cases, to be satisfactory from society's standpoint, technologies must be judged useful by farmers.

In most developing countries choices among alternative technologies are left to farmers. By now two related impressions about farmers are widely held: 1) farmers are purposive in their behavior, seeking to obtain incomes and to avoid risks; they are sensitive to the nuances of their environment; and they are reasonably efficient in managing the resources at their disposal and 2) while farmers' choices among alternative technologies are influenced by a host of variables, physical, biological, and economic forces dominate those choices.

This last impression warrants some amplification. Based on a series of CIMMYT sponsored country studies examining factors influencing the adoption of new maize and wheat technologies (essentially improved varieties and higher rates of fertilizer) it was concluded that:

> the most persuasive explanation of why some
> farmers don't adopt new varieties and fer-
> tilizer while others do is that the expected
> increase in yield for some farmers is small
> or nil, while for others it is significant,
> due to differences (sometimes subtle) in
> soils, climate, water availability or other
> biological factors (Perrin et al., 1976).

These studies and a reading of the earlier impressions of others (e.g., Foster 1962 and Schultz 1964) led to the conclusion that, while other variables might have a limited influence on choices among alternative technologies, income and risk are prominent farmer concerns and

these variables are strongly influenced by the natural
and economic circumstances of the farmers making the
choices. Hence, our emphasis is on these physical, biological, and economic factors.

With this view of farmers, technologies which will
be widely used must be consistent with farmers' natural
and economic circumstances and must promise improved incomes while keeping risks within reasonable bounds.
Technologies which do not meet these standards will not
be widely taken up.

The utility of technologies can also be judged from
the standpoint of a nation's goals. National decision
makers will want patterns of adoption to have consequences, e.g., for income distribution among producers or
for the distribution of benefits among consumers, which
are in accord with national goals. Given this concern,
those responsible for national policy will rarely be
indifferent about alternative technologies and, consequently, about alternative lines of research aimed at
forging improved technologies.

Procedures for Developing Useful Technologies

Orientation

Four points should be made before initiating a brief
description of our procedures for developing useful
technologies.

First, we are concentrating on that research whose
results are intended for near or intermediate term application, e.g., fertilizer research or plant breeding. We
are less concerned with basic or exploratory research
destined to be applicable in the long run.

Second, the entire process features collaborative
research among biological scientists and economists.
With farmers sensitive to both natural and economic
forces the formulation of technologies requires the same
sensitivity. This is not commonly found in a single
scientific discipline and even less in a single individual. In the partnership we envision the biological scientist contributes his knowledge of the interaction among
plants, insects, and diseases and their environment while
the economist brings an awareness of the influence on
farmer decision making of other opportunities for employ-
ing his resources and of markets for products and inputs.
Beyond this, for issues relevant to policy makers, biological scientists have clearer perceptions of what is
feasible through research while economists have the advantage in sorting out the implications of the adoption
of alternative technologies. Each, then, contributes
elements which are crucial in the formulation of technologies consistent with the needs of representative farmers
and with national goals. This collaboration is a hall-
mark of the procedures being described.

The third point is that we are concerned here with formulating technologies for a single crop or for that crop as part of a mixture. We are not discussing full-scale farming systems research.

Finally, the procedures aim at useful but not necessarily "optimal" technologies. After all, if each farmer responds to his own natural and economic circumstances then, as these differ among farmers, each could need a different "optimum" technology. Satisfying such demands is clearly beyond the capacity of any national research system. In place of "optimums" we seek to forge good approximations, technologies which promise more income with acceptable risks to representative farmers. We expect that, after adoption, each farmer will adjust the recommended practices to fit his own particular circumstances. This expectation is entirely consistent with experience, e.g., the increasing use of fertilizer on HYV wheats in India's Punjab and in Mexico's Yaqui Valley. Moreover, this stance relieves the researcher of the costly impression that he must be precise in framing recommendations. The researcher must be precise in his research, of course, but his recommendations are most useful when formulated as good approximations for a large number of potential users.

In brief, then, the procedures rest on collaborative research destined for early application, treat a single crop or mixture, and promise useful but not necessarily "optimal" technologies.

And there is one additional caveat. We recognize that the effectiveness of agricultural research is limited by shortages of physical and human capital, by nettlesome work rules, and by other constraints as well as by the limitations mentioned in our introduction. Even so, research is being done, technologies are being recommended, and farmers are following some recommendations but rejecting most. Hence, it is appropriate to question the paradigms which now organize applied research and it is potentially useful to explore new formats for the undertaking of such research.

Integrating Entities

A distinguishing feature of the process described in the following paragraphs is its emphasis on representative farmers as its primary clients. In our view, for many countries this represents a significant shift in the orientation of agricultural research. And what are the dimensions of this shift?

We believe that much agricultural research in developing countries is concentrated on problems emphasized by professional disciplines and guided by their standards. This is entirely consistent with the training of most active agricultural researchers and with the incentives which orient their efforts. It is also consistent with the paradigms followed in developed countries where technological change has contributed to rapid increases in yields and reductions in production costs.

Why, then, with the system featuring professional peers as primary clients apparently working so well in developing countries, shift the emphasis to farmers as primary clients?

Said briefly, we believe such a change will make agricultural research in developing countries even more effective. This conviction emerges from our interpretation of the process which links research to practice in developed countries. What is most emphasized in this process is the research of the publicly supported research systems. What is too little emphasized is the important role of entities which mediate between this research and the farmer, and which integrate research results into effective technologies for farmers.

These mediating entities, e.g., the agri-business complex in some countries, are not well established in developing countries. Moreover, unhappily, the incentives of developing country public institutions do not encourage the researcher to play an integrative role. On the contrary, incentives tend to accent professional contributions measured by the timely and lucid publication of research results, the contribution to professional organizations, and the training of others in the litany of the discipline. Furthermore, work rules seemingly conspire against anything done off experiment stations. The result is that research is often more attuned to the problems of the profession than to those of representative farmers and the recommendations are often irrelevant to their needs. It is the absence of this critical integrating activity which underlies our belief that there is scope for making research systems more effective.

We turn now to a brief description of the procedures we have been developing. Their function is to orient the competence of researchers toward the needs of farmers, bridging the gap between research and practice.

Identifying Relevant Farmers

Natural circumstances in most countries are usually sufficiently variable so that several technologies will be needed for a given crop or crop mixture. Moreover, farmers operating under essentially uniform natural circumstances might well confront such differing economic circumstances that they will need different technologies for a given crop. It is unlikely that the research resources of a country are sufficient to simultaneously meet all such demands, even for a single crop. The first step, then, in organizing research is to identify the farmers for whom technologies are to be formulated.

The process is expeditiously handled by grouping production areas into roughly homogeneous environments.

Within such an environment the crop or mixture in question reacts in roughly the same way and confronts roughly the same challenges. In producing areas assigned to other environments the crop mixture behaves differently in important ways. A first grouping can usually be made on the basis of the experience of informed biological scientists and economists working with secondary data on area, yield, soils, weather, elevations, and demography, all complemented by the observation of merchants specializing in the crop.

The next step is to roughly characterize the environments in terms of information which may be important to agricultural policy, e.g., area in the crop, production, number of farmers, distribution of farm size, relative importance of the crop, and exportable surpluses. Combining this information with researchers' impressions of the potential for improving technologies is usually sufficient to permit a first rough ordering of the environments in terms of national goals.

In Ecuador this procedure was followed to identify five environments in which farmers produce maize. It was inferred from policy statements that government was emphasizing the incomes of low income farmers. For each environment the area in maize, maize as proportion of total cropland, average farm size, and yields were estimated. Happily, the zone with the smallest farms and the heaviest reliance on maize also had one of the largest areas in maize and biological scientists ranked it high in terms of the potential for forging improved technologies. This congruence will not always occur, so the rankings will often have a degree of arbitrariness, becoming more so as government goals are less clearly stated and as impressions about research potential are more probabilistic.[1]

Identifying Farmers' Circumstances

While secondary data are adequate to frame general impressions, they are rarely sufficiently detailed to orient research on improved technology. Such detail requires firsthand knowledge of circumstances and problems. We advocate two related sets of activities for acquiring that firsthand information. Again, given the scarcity of research resources, these are concentrated on the environments assigned the highest priorities.

The first of the activities is exploratory survey work in the environments for which technology is to be

---

[1] For the most part, examples are from Latin America. We could also have taken them from CIMMYT work in East Africa or South Asia.

developed. This will include informal but organized discussions with farmers, with merchants, and with others familiar with the environment. The effort involves both discussion and observation and focuses on production practices and problems, markets for production and inputs, and important competing activities.

Secondary data, the knowledge of researchers, and the results of the exploratory survey are then used to describe tentative recommendation domains (i.e., sets of farmers whose natural and economic circumstances are sufficiently similar that a given technology will be relevant to each farmer within a set).[2]

The second activity starts with the same sources of information plus the insights derived from the exploratory survey and proceeds to a formal survey. The information and insights are integrated into questionnaires, which are then administered to a random sample of farmers from each tentative recommendation domain. While each questionnaire is focused on issues critical to the farmer, the farm, and to the crop or crop mixture of primary concern, it also deals with other activities--other crops, livestock, non-agricultural activities, or non-farm activities--which impinge in important ways on the crop or mixture under study.

These surveys, especially the formal survey, serve to identify characteristics of representative farmers, e.g., farm size, common implements, typical rotations, critical periods, and access to inputs. They permit description of practices currently employed--levels, types, and dates associated with each activity--by representative farmers. They provide information for establishing the representative farmer's perception of major problems affecting the crop or mixture under study. Finally, the survey data also allow for refinement of the description of recommendation domains.

The procedure starts, then, by grouping farmers into essentially homogeneous natural environments, orders these environments in terms of national goals, assesses farmers' circumstances, establishes groups of farmers in terms of natural and economic characteristics and national goals, and makes specific the circumstances of representative farmers for each important group.

Returning to the example of Ecuador, surveys there indicated that the environment assigned the highest priority contained three different sets of farmers based on natural factors. The three emerged from insect patterns and access to irrigation. The insect patterns, in

---

2 While some of our colleagues find other phrases more congenial, we favor this one. Notice that adjacent farmers need not be in the same domain and that recommendation domains need not be contiguous in space.

turn, were closely related to altitude. Some differences
in economic circumstances appeared, e.g., farm size and
access to inputs; for virtually all farmers in each group
these differences were slight. The remaining farmers
were few in number and small in the proportion of total
area given over to maize. So, no additional recommendation domains were formed because of economic circumstances. For each domain the survey data were used to
characterize the circumstances of the representative
farmer.

While data on farmer circumstances are gathered
primarily to orient research, experience shows that an
immediate sifting of the information for policy implications might also be profitable. For example, one maize
study showed that a supposedly effective system for dis-
tributing inputs was falling far short of meeting farmer
requirements for insecticides. The problem uncovered,
policy makers could move to clear it up.

The perceptions of farmers and merchants, the knowledge of scientists, and the information derived from surveys are then combined to reveal factors significantly
limiting the production of representative farmers. As
with the earlier activities, data analysis requires the
joint participation of biological scientists and economists. Each, again, brings specialized skills and sen-
sitivities to the data, contributing to the identification of significant problems and to establishing the
lines of work which might lead to their resolution. The
research itself is undertaken on experiment stations and
on the fields of representative farmers (Fig. 1).

Organizing Experimentation

Some of the limitations identified require research
under carefully controlled conditions. This is usually
best done on experiment stations. Its benefits often
will not be realized in the near term and its results
must be tested under the conditions of relevant representative farmers.

The surveys also orient on-farm experimentation.
The first step involves examining existing solutions to
the problems identified, carefully assessing the adequacy
of such solutions, and modifying proposed solutions in
the light of findings on the fields of representative
farmers. This activity has a featured role in the process because the natural conditions of experiment stations often depart markedly from those of representative
farmers.

Survey work in one Andean region of Peru showed the
importance of leaf diseases in maize. The importance of
the diseases established, maize breeders began to screen
their own material and sought promising materials from
others to screen for resistance to this disease. In
another Andean region survey work uncovered a farmer

Fig. 1. Overview of an integrated on-farm research program.

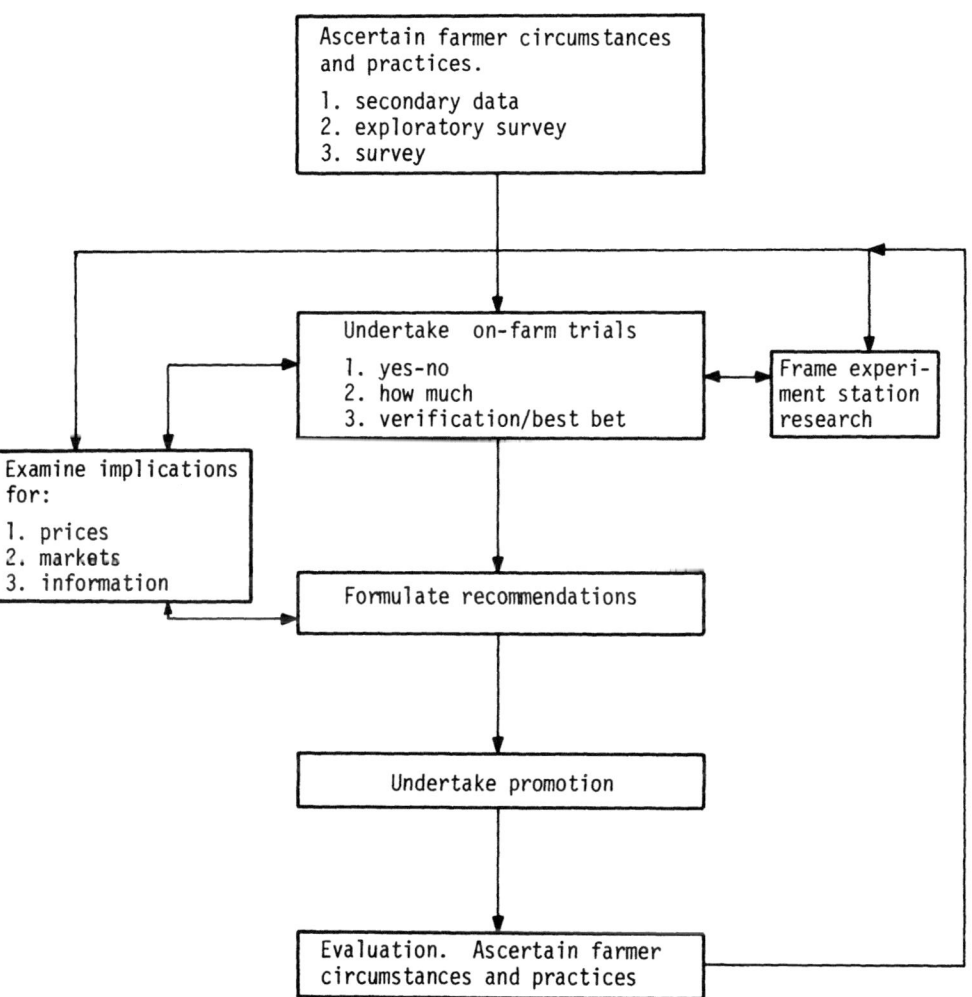

demand for a shorter season variety with good stalk
strength. Maize breeders are now recombining shorter
season material with material having good stalk strength
and proper grain type. And why good stalk strength? Because surveys disclosed that the representative farmer
grows climbing beans with his maize and on-farm experiments showed that existing short season varieties were
unable to carry the weight of the beans. These problems
and opportunities were uncovered through on-farm research
involving surveys and experimentation.

On-Farm Experimentation

The on-farm trials are initiated with best-bet strategies based on the experience of researchers and farmers'
perceptions. At each critical period in the life of the
crop or mixture farmers and researchers come together
around the crop to assess the adequacy of the strategies.
Information from the trials flows to the experiment station, signaling new problems, and to trials in succeeding
years (Fig. 2). Each year information from experiment
station trials is assessed for its relevance to the problems judged most critical.

Three classes of on-farm trials are advocated: yes-
no trials, how much trials, and verification trials. The
yes-no trials are designed to look at major effects and
first order interactions of the factors thought to be
most critical in limiting production. Factorial designs
are the mainstay of these trials and these feature two
levels of the inputs or practices being examined, one at
current farmer levels and the other at a significantly
higher level. The how-much trials are designed to identify levels at which income seeking, risk averting farmers might want to employ inputs or practices detected as
limiting in the yes-no trials.

In developing improved technologies there are always
questions regarding how many factors can be changed at
one time, to what degree input use can be changed, and at
what level those factors not being changed should be set.
For on-farm experiments, we advocate that attention be
concentrated on only three or four factors at a time.
Most evidence is that farmers tend to make but a few
changes at a time, concentrating on those with the highest payoffs. So, research can be concentrated on a
limited number of factors rather than aiming at all
potential changes in one fell swoop. Regarding the
levels of input use, profit and risk considerations require that rates of return on purchased inputs be quite
high, i.e., probably higher than the apparent cost of
capital, and this could suggest less intensive use than
might be thought desirable by yield maximizing biologists
or profit maximizing economists. How much less can be
approximated with farmers during research and verification trials.

Fig. 2: On-farm trials under farmer circumstances.

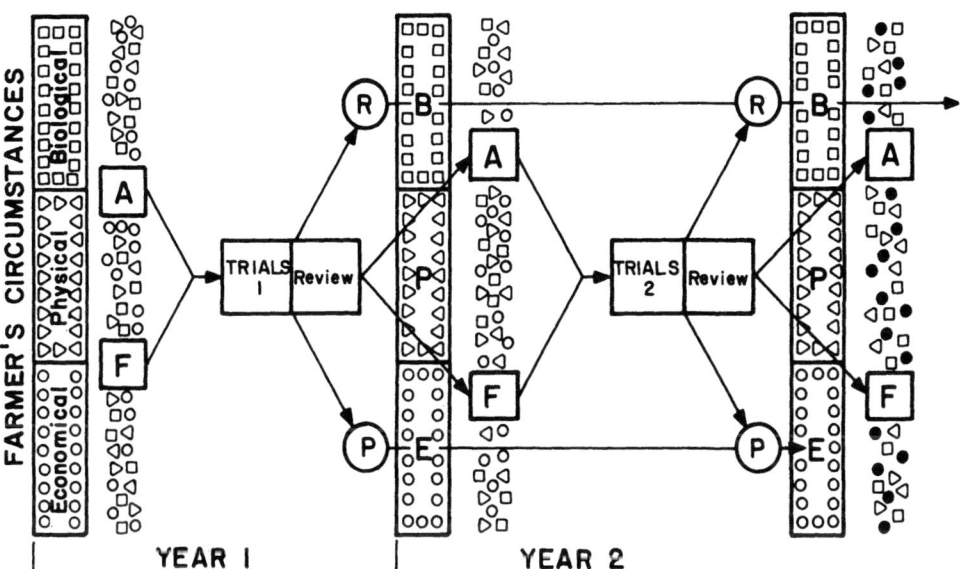

In the first phase the farmer (F) and the research team (A) come together
in the farmer's environment, ascertain important problems, and identify
potential solutions. These are tried out as "Best Bets" in a first set
of on-farm experiments. The trials are monitored by (F) and (A). (A)
and (F) use resulting information to adjust subsequent trials. Information goes also to station researchers (R) and to policy makers (P),
who organize their work to alter the farmers' environment. (This is
exemplified by a change in the economic circumstances after Year Two,
e.g., different prices, giving rise to a new environment in Year 3.)
Interaction continues until a technology judged suitable for verification
is identified.

Finally, we believe that the nonexperimental factors, those not part of the yes-no trials, are best set to match practices followed by representative farmers. By definition these variables are not important in determining yields or costs--else they would be among the experimental variables--so they can be set at low cost rather than at high cost levels.[3]

Each year best-bet strategies are reformulated in terms of the on-farm trials of the previous year and the impressions of all participants in the trials (Figs. 1 and 2). They are also modified to incorporate findings from experiment station research. Once farmers and researchers are convinced that an appropriate strategy is available, i.e., one consistent with farmers' circumstances and promising significant improvement in income at acceptable risk, the strategy is verified on a larger number of representative sites. Once verified, recommendations are made.

Notice that the process accents immediacy with improved technologies available in the near or immediate term. If all goes well--if the proper elements have been integrated in the research--the recommended technologies will be widely and rapidly diffused. This occurs precisely because they have been deliberately tailored to fit the needs of representative farmers.

Over time, individual farmers will adjust the recommendations in the light of their particular circumstances. Experiment station results, e.g., new varieties, will be available for testing under farmers' circumstances and incorporated in new best-bet strategies. In the longer run, researchers will turn their attention to other environments or to other problems of lesser importance in the same environment. The process, then provides for continuing improvement in recommended technologies as both farmers and researchers, from on-farm trials and from experiment stations, apply new experience and information to farmer problems.

Incentives and Structure

The process described here rests squarely on bringing publicly sponsored researchers together around the

---

[3] Some CIMMYT staff members hold a different view on this point. Largely because of their contention that anything done on farmers' fields is regarded by farmers as a demonstration, they advocate setting nonexperimental variables at levels sufficiently high that the expression of experimental variables is not limited.

problems of representative farmers. By basing research
on representative natural and economic circumstances, researchers will play that important integrative role. In
many cases implementing such research will require
changes in incentives and in work rules. For at least
some researchers, incentives must favor contributions to
representative farmers and to production; work rules must
facilitate on-farm efforts.

Before making these changes, of course, the utility
of the procedures themselves must be demonstrated. We
believe that favorable evidence is accumulating rapidly.
Already several national programs are recasting research
in terms of the earlier discussion and their on-farm
activities are showing new solutions for the problems of
representative farmers. These solutions are moving towards verification. We have yet to see whether they give
rise to recommendations suitable for the target groups of
farmers but we are optimistic about developments.

Summary

The preceeding paragraphs describe a procedure for
developing improved technologies. Farmers are at its
core as its primary clients. The procedure focuses on
ascertaining relevant farmer circumstances and integrating these into research aimed at developing improved
technologies. It rests on collaboration among farmers,
biological scientists, and economists so that the special
experience and skill of each can influence the orientation of research. On-farm research, under the circumstances of representative farmers and with feedback from
year to year and experiment station research, plays a
featured role. The process itself is "non-perfectabilitarian"; it does not envision developing "perfect" technologies. Rather, it systematically focuses on major
constraints to production, integrates natural and economic circumstances of representative farms, provides for
continuing and immediate improvement through research,
and counts on individual farmers to make adjustments in
terms of their own special circumstances.

References

Foster, George. 1962. Traditional cultures, and the
impact of technological change. Harper and Row,
New York.
Perrin, Richard and Donald Winkelmann. 1976. Impedi-
ments to technical progress on small versus large
farms. American Journal of Agricultural Economics,
Vol. 58, No. 5.
Schultz, T. W. 1964. Transforming traditional agriculture. Yale University Press, New Haven, Connecticut.

4
An Ecological Systems Conceptual Framework for Agricultural Research and Development

Robert D. Hart

Traditional agricultural disciplines have evolved by dividing the agricultural production process into smaller and smaller units. Some of these divisions are structural, such as the separation of plants and animals, while others, such as the difference between the disciplines of physiology and economics, are based on functional characteristics. When multidisciplinary teams are formed to study a unit that encompasses structural and functional components and processes that have traditionally been assigned to different disciplines, integration of the team is often hindered by the lack of a common conceptual framework.

A conceptual framework must be more than a set of definitions agreed upon by a multidisciplinary team. The framework should function as an integrative tool that allows all team members to understand the relationship between disciplines, as well as the relationship between specific disciplines and the larger unit that is the subject of study.

The decision to form a multidisciplinary team often occurs after it has been demonstrated that different disciplines working separately have been less successful than expected. This has occurred in tropical agricultural research and development programs. Research scientists have recently recognized the necessity of working with units larger than the individual crop or with specific processes such as economic transactions. As a result, cropping system and farming system multidisciplinary teams are being formed in many tropical agricultural research institutions. The conceptual frameworks used by existing teams have usually developed by an evolutionary process as the team attempts to conceptualize the unit being studied and integrate different disciplines.

The primary purpose of this paper is to describe a general agricultural systems conceptual framework that can serve as a starting point for a multidisciplinary team. A conceptual framework is a model, and like any model, represents a simplification of reality.

Simplification involves assumptions, which in effect are hypotheses as to the structure and function of the unit under study. The validity of these assumptions and the potential of the conceptual framework can best be evaluated by applying the model to reality and analyzing the results. In this paper I describe an ecological systems conceptual framework and apply this model to the reality of the agricultural production process of Central American small farmers.

The Ecological Systems Model

A system is an arrangement of components that function as a unit. Biological and physical systems are open systems, i.e., they interact with their environments, processing inputs to produce outputs. The systems approach was pioneered in biology by Smuts with his introduction of the concept of holism in 1926 (Becht, 1974). In the early 1930's von Bertalanffy (1968) formulated what he defined as a General Systems Theory.

The systems approach has been applied to all biological disciplines, but is probably most associated with ecology. In 1935 Tansley proposed the term ecosystem (Evans, 1956). The concept has been developed by many others, such as in the classic papers on trophic levels by Lindeman (1942) and energy flow through ecosystems by H. T. Odum (1957). Development of the ecosystem concept into a larger ecological systems concept is probably most associated with E. P. Odum (1971) and his *Fundamentals of Ecology* text and the energy circuit approach of H. T. Odum (1971).

E. P. Odum defines an ecosystem as "any unit that includes all the organisms in a given area interacting with the physical environment so that a flow of energy leads to clearly defined trophic structure, biotic diversity, and material cycles within the system." The flow of energy and cycling of materials associated with ecosystems can be found in other ecological systems both larger and smaller than ecosystems. In systems terminology, ecosystems are subsystems of other systems as well as composed of subsystems. The conceptual framework of ecology is based on the assumption that there exists a series of hierarchically interacting systems from the universe to the smallest subatomic particle.

Ecological studies are usually applied to only one or two levels of the universe-to-subatomic particle hierarchy. Ecosystems, communities, and populations are probably the most common units studied in ecology. Each hierarchical level is conceptualized as a system composed of a set of subsystems. Interaction between two subsystems of the same system can be defined as horizontal system interaction. Horizontal system interaction can be superimposed upon the vertical system interaction implied by the universe-to-subatomic hierarchy. This vertical and horizontal ecological systems model can also be applied to the agricultural production process.

Hierarchial Agricultural Systems

If the hierarchial ecological systems conceptual framework is applied to an agricultural production process, a set of hierarchically related agricultural systems emerge (Fig. 1). As in the case of the ecological systems framework, agricultural systems exhibit not only vertical hierarchical system interaction, but also horizontal system interaction. Each hierarchical level is a functioning set of subsystems with the outputs of some subsystems acting as inputs to others. While it is possible to describe a global level agricultural system, from the point of view of agricultural research and development, the geographic region is probably the largest unit of interest.

A regional agricultural system includes all the farms in the geographic region; the marketing, credit, and information centers; and the infrastructure that ties these regional subsystems together. A region can be analyzed as a system with materials, energy, money, and information flowing into and out of the region and between subsystems within the region. From an agricultural research point of view, the farms within the region are the most important subsystems and form the next lower hierarchical level under the region.

A farm is also a system made up of subsystems. A farm system can be viewed conceptually as a set of spatially definable areas in which either crops, animals, or both are produced, and a homestead area where the farm house is located. The crop or animal production areas form units, analogous to the ecosystem unit in ecology, and can be defined as agroecosystems. The farm house area in which the farm family is fed and clothed and the economic transactions and management decisions that occur on a farm can be combined to form a socio-economic subsystem of the farm system. The socio-economic subsystem and the agroecosystems interact to form a farm system. If agricultural research is of primary concern, the agroecosystems of a farm system are the most logical next lower hierarchical level to be analyzed in more detail.

An agroecosystem is also a system made up of subsystems. As in the case of natural ecosystems, it is composed of a biotic community of plants, animals and micro-organisms and the physical environment in which the community functions. Energy flows between trophic levels and materials are cycled. An agroecosystem differs from a natural ecosystem in that at least one plant or animal population is of agricultural value and that man plays an important management role. Soils, crops, weeds, insects, and micro-organisms can be defined as subsystems of crop-dominated agroecosystems. In a domesticated animal-

Fig. 1. Hierarchical relationship between agricultural systems.

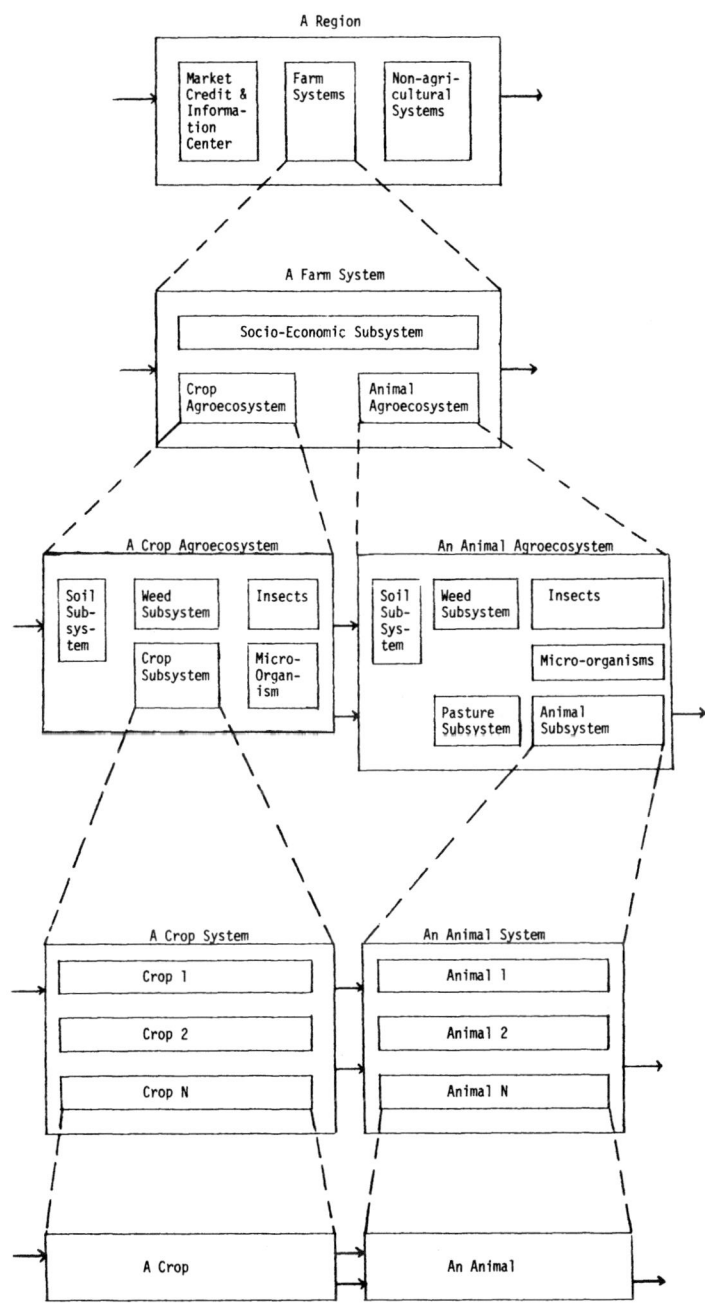

dominated agroecosystem, soils, pasture, weeds, insects, micro-organisms, and domesticated animals make up the subsystems that function as a unit in the agroecosystem. Agronomic research has been done on all of these sub-systems, but crop systems and animal systems have receiv-ed the most attention.

A crop system is an arrangement of crop populations that process energy (solar radiation) and material inputs (soil nutrients, water) to produce outputs (crop yield). The crop population can be arranged both spatially (plant-ing distances) and chronologically (date of planting). When more than one crop species are combined in space and time, the resulting assemblage can be exceedingly complex. The individual crop species are subsystems of the crop system and make up the next hierarchical level under the crop system. The individual crops can also be subdivided into hierarchically lower subsystems as physiological processes. In agronomy considerable attention has been given to this hierarchical level with the recent emphasis on the study of crop architecture and crop genetic sys-tems as part of crop breeding programs.

A domesticated animal system is an arrangement of animal populations that processes energy and material in-puts (pasture, feed supplements, etc.) to produce outputs (meat or animal products). An animal system is on the same hierarchical level as a crop system. Animal popula-tions made up of individual animals composed of inter-related physiological systems form the next lower hier-archical level.

In applying the agricultural systems conceptual framework to a specific case, it is not always necessary or practical to use the entire hierarchy. Emphasis can be placed at one level, as for example in the case of a cropping systems project. In principal, however, it will always be necessary to study at least three levels: the unit of interest, the next higher level, and the next lower level. The next higher system must be studied in order to measure the inputs into the system, and the next lower level must be studied in order to understand how the system functions. In the case of a cropping systems project, activities will need to be applied to the agro-ecosystem, to the crop system, and to crop levels. A farming system project must study regions, farm systems, and agroecosystems.

The first step in a region, farm agroecosystem, or crop or animal system study is the construction of a qualitative model of the unit under consideration. In the context of this framework, model building involves identifying the inputs and outputs of the system of interest, the subsystems of the system, and the circuitry connecting these subsystems. The next step is to begin to quantify the relationships hypothesized in the qual-itative model, and to construct a quantitative model of the system. The precision required depends upon how the model will be used.

The qualitative models that would be developed by a multidisciplinary team if the hierarchical agricultural systems model were used, would vary with the ecological and socio-economic conditions of a specific region, farm, agroecosystem, or crop or animal system. However, these systems have general inherent characteristics that make it possible to outline general qualitative models for each level of the hierarchy. I have assumed that these models would be used for research and development purposes.

Regional Development

Figure 2 is a qualitative model of a regional system. The major inputs and outputs into a region can be classified into energy, materials, money, and information. The first step in any regional study would be to identify these inputs and outputs. Energy, materials, money, and information also flow between the subsystems of a region. In the model, agricultural subsystems of a region are defined as market, credit and information centers, and the different types of farm systems within the region. In a specific regional development study these farm systems would be identified and classified. This same information would, of course, also be necessary for a farm system study, since the first step in a farm system study would be the selection of a specific farm system type, and the region would have to be studied in order to identify the inputs and outputs into the farm system.

Farm System Research

Figure 3 is a qualitative model of a farm system. The farm is divided into a socio-economic subsystem and agroecosystems. The inputs and outputs into a farm system can also be classified into energy, materials, money, and information. The inputs and outputs into the agroecosystems of the farm system can be grouped into information, energy, and materials categories. Information enters an agroecosystem in the sense that human, animal, or machine energy enters an agroecosystem as part of a management plan. Farm system research requires not only the construction of a qualitative model describing these relationships, but also a quantitative model where real numbers are assigned to the farm system inputs and outputs and the flows between the subsystems of the farm. The primary objective of farm system research would be to use the model to identify possible modifications of an existing farm system or to design a new farm system. The constraints upon this design process, such as labor availability, nutrition requirements of the family, etc. would be determined before the generation of a new farm system. The regional system and the socio-economic

Fig. 2. Flow of money, materials, energy, and information through a geographic region.

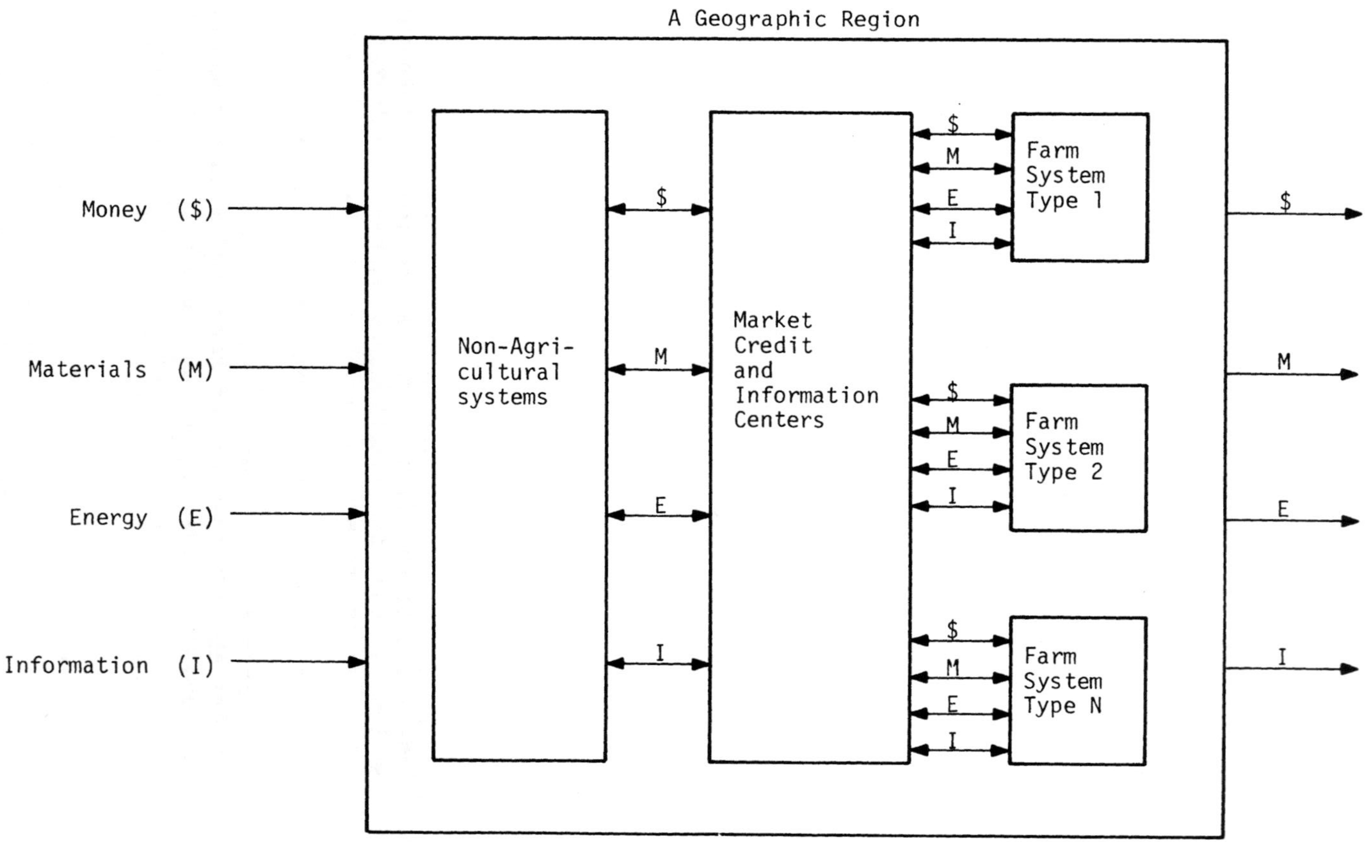

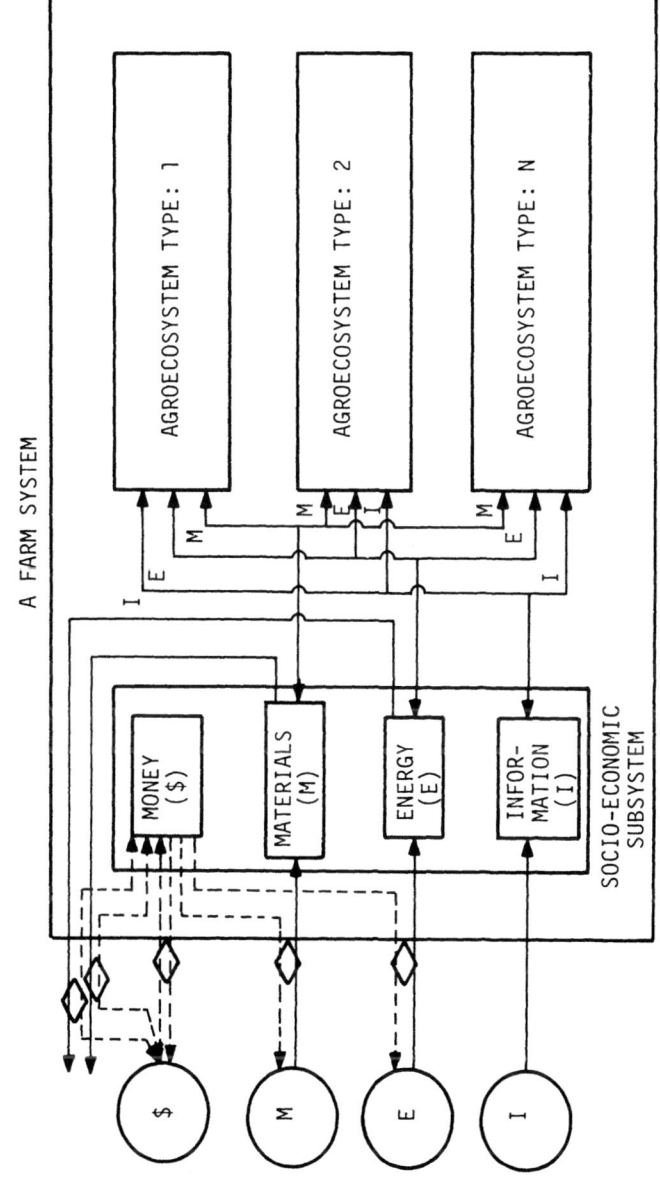

Fig. 3. Flow of money, materials, energy, and information through a farm system.

subsystem of the farm would be studied to identify socio-
economic constraints, and the agroecosystems would be
studied to identify the physical and biological constraints.

Agroecosystem Research

Figure 4 is a qualitative model of a crop agroecosystem. In the diagram, physical and biological sources
of inputs such as solar radiation, crop seed, herbivores,
etc. are shown entering the system on the left side;
agricultural chemicals such as fertilizer, herbicide,
etc. enter from the bottom; and human, animal, and
machine energy enter from the top as determined by an
agroecosystem management plan. The agroecosystem is an
extremely important research unit, primarily because it
is the unit that the farmer manages. While the performance of the crop system within the agroecosystem is the
key to agricultural production, this performance is
regulated by managing the agroecosystem. The Agroecosystem Management Plan is a convenient information package
for transferring alternative technology to a farmer.
The subsystems of crop agroecosystems are soil, weed,
herbivore, micro-organism, and crop systems. Water and
nutrients are outputs of the soil system and, along with
solar radiation, form potential inputs that are competed
for by crops and weeds. Crops and weeds process these
inputs to produce biomass that is an input to herbivores
and micro-organisms that in turn recycle nutrients to the
soil subsystem for subsequent uptake by crops and weeds.
As in any ecosystem, the cycling of materials is powered
by a flow of energy through the system. From an agronomic perspective, the output of economic crop biomass
(yield) is the most important output from the system.
Agroecosystem research has the ultimate objective of
modifying either the management of the agroecosystem, the
crop system, or both. Research with this objective will
require experiments with analytical objectives in order
to understand how the system functions (build qualitative
and sometimes quantitative models), as well as experiments comparing potential modifications with existing
agroecosystems in specific areas.
Figure 5 is a qualitative model of an animal agroecosystem. Ecologically, animal agroecosystems and crop
agroecosystems are very similar. In animal agroecosystems, natural herbivores are replaced by domesticated
animals and pasture is substituted for natural plants,
while in a crop agroecosystem only the natural plants are
replaced. This substitution is not 100 percent effective,
and weeds and natural herbivores are still part of agroecosystems. Animal and crop agroecosystems are sufficiently similar so that the same methodology suggested
for crop agroecosystems also applies to animal agroecosystems. Animal agroecosystems can be improved by

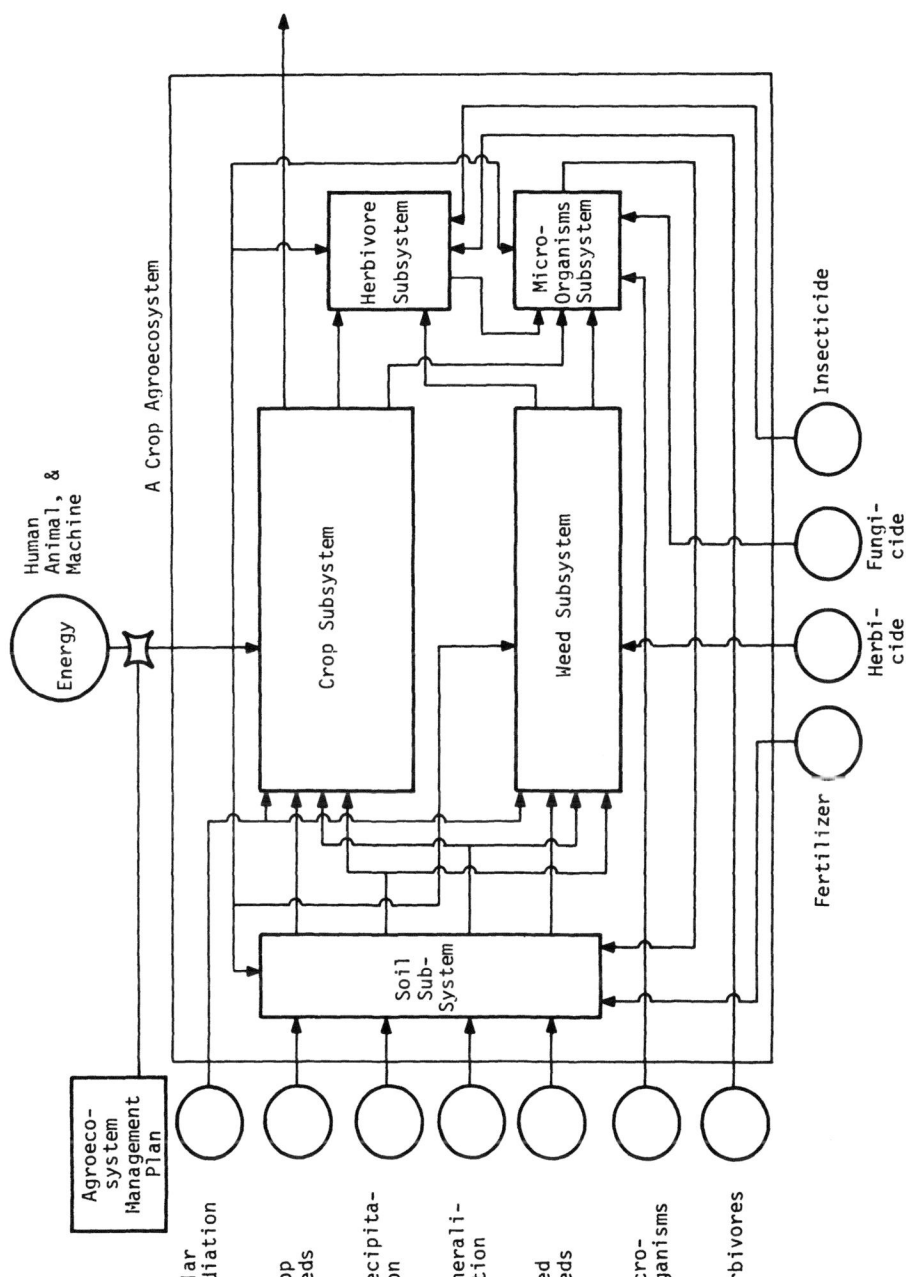

Fig. 4. Flow of materials and energy through an agroecosystem with a crop subsystem.

Fig. 5. Flow of materials and energy through an agroecosystem with an animal subsystem.

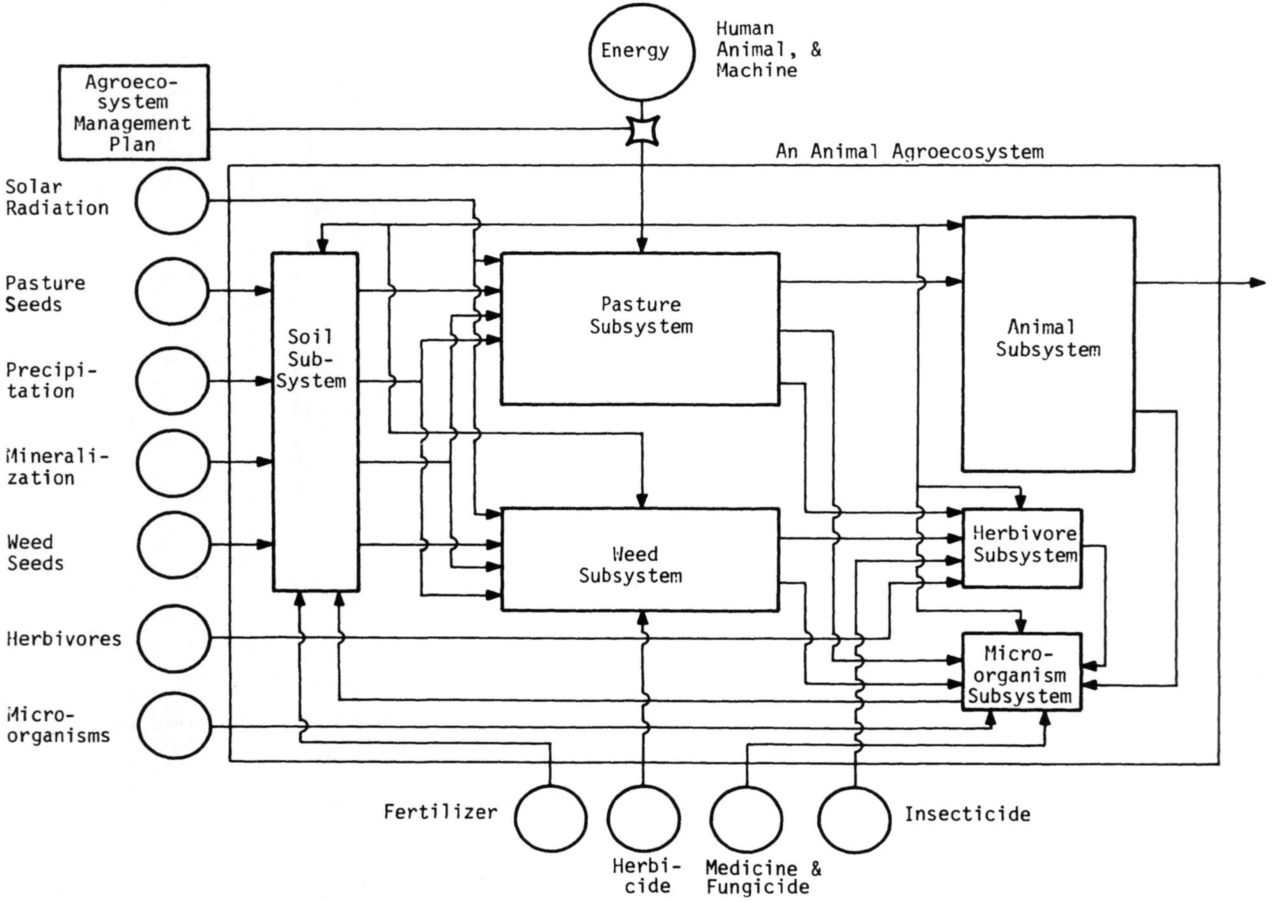

55

modifying the management and inputs into the agroecosystem or modifying the animal subsystem.

Crop System and Animal System Research

     Figure 6 is a qualitative model of a crop system and
an animal system. Crop or animal system research need
not always be done while the systems function as subsystems of an agroecosystem, but research at the agroecosystem level will definitely be necessary to define the crop
or animal system properties that can be studied in iso-
lation.
     A crop system can be modified by changing the spatial
arrangement between crop populations (planting distances),
the chronological arrangement of the crop populations
(time of planting), or the crop components (either variety or species) of the system, or any combination of
these modifications. One crop can be substituted for
another (substitution), crop species can be added (diversification), or substitution, diversification, and intensification can be combined.
     In the crop system diagram (Fig. 6), crop popula-
tions (1, 2, - N) are arranged with a space x time
dimension. This crop arrangement forms a pattern and is
sometimes defined as a cropping pattern. Ideal cropping
patterns are determined by input functions (e.g., rainfall distribution) and the available crop components. If
these input functions and available genetic material
remain constant for a sufficient length of time, farmers
usually evolve cropping patterns that are in equilibrium
with these constraints. Unless new varieties of crops or
new inputs are made available, it is highly unlikely that
a better cropping pattern can be found than the pattern
already evolved by farmers.
     Crop system research can have short-term objectives
such as the identification of better crop systems through
a trial and error approach of comparing potential systems
with the farmer's system, or long-term objectives such as
the identification of crop system design principles and
an understanding of how crop systems function. The long-
term objectives are, of course, only long term in the
sense that the period of time before the first practical
recommendation is available will be quite long; ultimately, an understanding of how the system functions is the
fastest way to produce viable recommendations.
     Animal systems are spatial and chronological
arrangements of animal populations. The animal populations in an animal system usually consist of different
age and sex classes of the same species, although in some
cases different species, such as pigs and chickens,
occupy the same space and compete for some of the same
resources. In the animal system diagram (Fig. 6) the
space dimension is divided into N subareas. Some
animal populations are rotated between subareas. Others

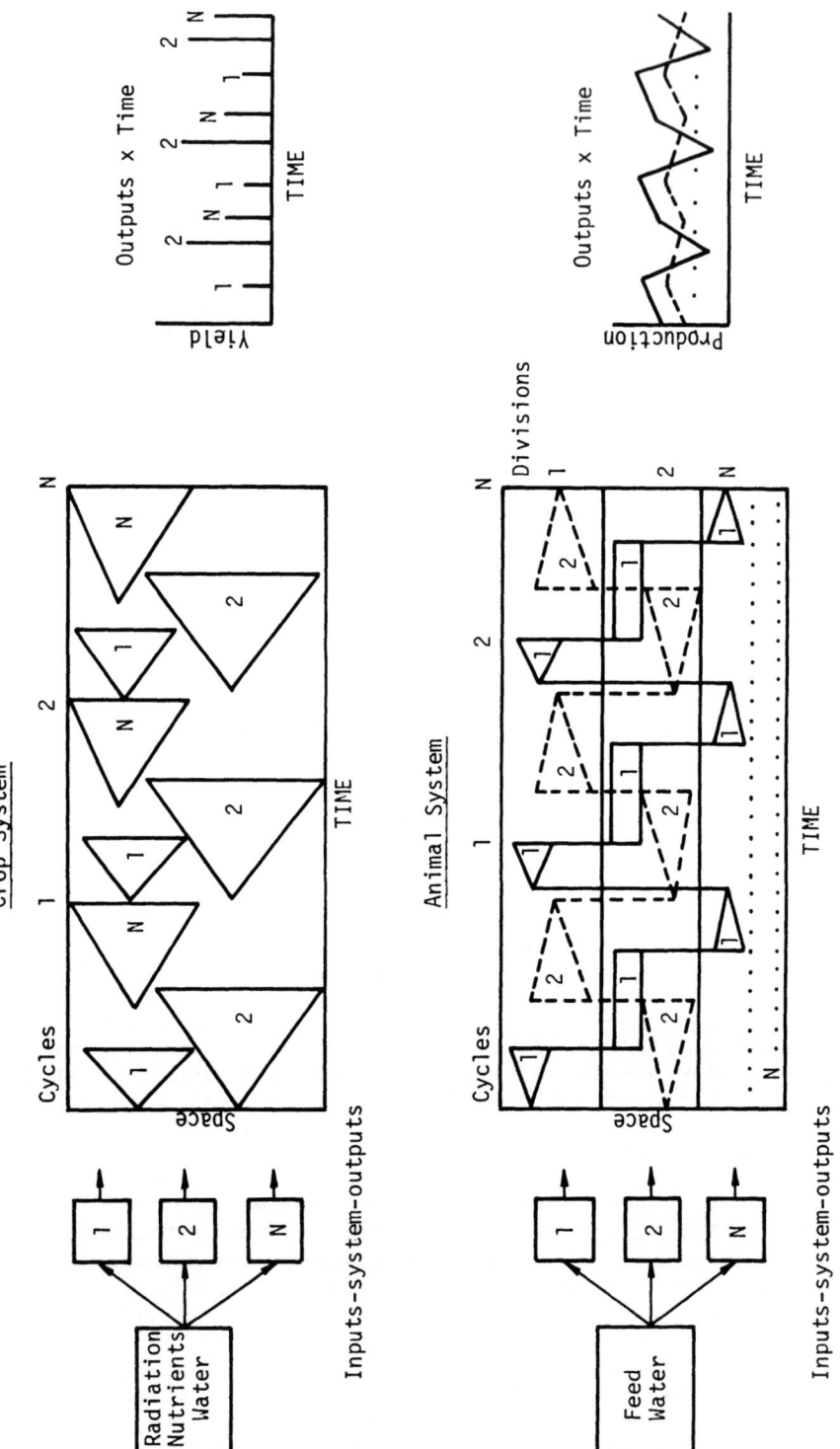

Fig. 6. Crop and animal systems as spatial and chronological arrangements of crop and animal populations, respectively.

are confined in one area. All animal populations interact in either space or time with at least one other population. The arrangement of animal populations forms a
pattern analogous to the cropping pattern of a crop
system.

An animal system can be modified by changing the
spatial or chronological arrangement of the animal population, the animal components of the system, or both.
The animal populations can change through substitution,
diversification, intensification, or combinations of
these modifications.

The Agricultural Systems Framework as a Team Integrator

Traditional agricultural disciplines can be divided
into horizontal one-level disciplines and vertical disciplines that cross hierarchical levels. Examples of the
former are crop genetics, soil fertility, and entomology.
Economics and ecology are examples of vertical disciplines. Economics concentrates primarily on vertical
relationships such as the chain from the farm to the
market to the consumers, while ecology encompasses both
vertical and horizontal systems relationships.

A multidisciplinary team should include both vertical and horizontal disciplines. If the entire agricultural system hierarchy from a region to the crop or
animal level is under study, integration of the regional
and farm systems study can probably best be done by an
economist, as almost all flows of energy and materials
between these levels are associated with a flow of money.
Farm system to crop or animal integration should be done
by an ecologist since the interaction of physical and
biological factors dominate these systems. Horizontal-
discipline scientists should be assigned responsibilities
within hierarchical levels. If the methodology of first
building qualitative models and then proceeding to quantify relationships is followed, all disciplines should
concur that the qualitative model represents a first
approximation of reality. Different disciplines can then
be assigned the responsibility of quantifying different
qualitative relationships.

The agricultural systems hierarchical conceptual
model described in this paper is only a preliminary
framework for a multidisciplinary team. As relationships
between systems are better understood, the conceptual
framework will need to be modified to reflect the characteristics of the phenomenon under study.

References

Becht, G. 1974. Systems theory, the key to holism and
    reductionism. Bioscience 24(10): 569-579.
Evans, F. C. 1956. Ecosystems as the basic unit in
    ecology. Science 123: 1127-1128.

58

Lindeman, R. L. 1942. The trophic-dynamic aspect of
    ecology. Ecology 23: 399-418.
Odum, E. P. 1971. Fundamentals of ecology. Saunders,
    Washington, D. C. 574p.
Odum, H. T. 1957. Trophic structure and productivity of
    Silver Springs, Florida. Ecological Monographs 27:
    55-112.
Odum, H. T. 1971. Environment, power, and society.
    Wiley, New York. 331p.
von Bertalanffy, L. 1968. General systems theory.
    George Braziller, Inc., New York. 295p.

5
One Farm System in Honduras:
A Case Study in
Farm Systems Research

Robert D. Hart

Agricultural scientists have recently recognized that farmers in tropical environments often plant crops in such a way that interaction occurs between crop species. These multi-species crop systems are presently being studied by many national and international research institutions. The success of these programs has demonstrated the potential of doing research with units larger than the individual crop.

One of the reasons crop systems research programs have been successful may be that the research is directed towards a unit that is consistent with a unit managed by farmers and the technology generated by the research programs can be directly adopted by farmers. This is not the case with crop-specific research results. The farmer has to integrate the crop-specific technology into his crop system before he can adopt it.

If consistency between the unit managed by farmers and the unit studied in agricultural research programs is important to the successful adoption of new technology, the study of whole farms (the largest unit managed by a farmer) would seem to offer great potential. However, farms are complex agricultural systems. Interaction may occur not only between crops and between animals, but also between crop systems and animal systems. At present, farm systems research is still in a conceptual and methodology development stage.

The farm system case study summarized in this paper was part of a crop systems research project conducted at Yojoa, Honduras between 1976 and 1979. Since farm systems form the environment in which crop systems function, one of the objectives of the study was to describe the structure and function of a dominant farm system in the Yojoa area and to use this information as a guideline for the crop systems research. Another important objective was to evaluate the concepts and methodology used. Although this paper includes a summary of the data collected, this information is presented primarily to illustrate the concepts and methods used in the study.

59

Methods

Yojoa, Honduras is a small village with approximately 200 farm families. The average farm size is eight hectares, but the most frequent farm size is between three and five hectares. The Yojoa area is approximately 100 meters above sea level with 1500 mm annual rainfall distributed in a bimodal pattern and with rainfall peaks in June and September. Very little rainfall occurs between February and May. Crops are usually planted in June and November. Maize, rice, and beans are the most important crops in the area.

In February 1976, a survey was conducted with the primary objective of identifying and describing the most important crop systems in the area. General socioeconomic data were also collected. The results of the survey were used to describe a representative farm, and a local extension agent was asked to identify five farmers meeting these criteria. The farmers were interviewed and Mr. Aureliano Alvarado was chosen for the case study.

A questionnaire (outlined in Table 1) was designed on the basis of a qualitative farm system model (Fig. 1). In the model, a farm system was conceptualized as a system with a socio-economic subsystem (the house and all social and economic components) and one or more agroecosystems (a crop system and the soils, weeds, insects, and diseases that interact with it).

The farm system was assumed to have inputs and outputs of money, materials, energy, and information. Money (shown as a dotted line) always flows in an opposite direction to materials and energy. For example, if a farmer buys fertilizer, materials flow in and money (what the farmer pays) flows out. If the farmer sells maize, materials flow out and money (what the farmer receives) flows in. The model also includes the possibility of money buying money, as when a farmer pays interest for credit.

Materials, energy, and information also flow between the socio-economic subsystem and the agroecosystems and between the agroecosystems. Money was not included as a flow between the subsystems of the farm system since economic transactions were assumed to occur only on the farm level and not within the subsystems of the farm.

Beginning on May 31, 1976, each week for one year Mr. Alvarado was interviewed and the questionnaire was filled out. At the end of 52 weeks, the weekly interviews were terminated and the data analyzed. The qualitative model (Fig. 1) was modified to include the agroecosystems and the flows of money, material, energy, and information identified during the study; the one-year totals for these flows were calculated; and a quantitative model (diagram) was drawn. Each flow was inspected to see if it was static (low weekly variability) or dynamic (high weekly variability). Dynamic flows were

Table 1. An outline of the questionnaire used in a
          farm systems case study at Yojoa, Honduras.
          1976-1977.

_____

  I. Farm System Input - Output

      A. Output of money

          1. crop-related expenses

          2. animal-related expenses

          3. household expenses

          4. others (debts, gifts, trips, etc.)

      B. Input of money

          1. crops sold

          2. animals and animal products sold

          3. off-farm family labor

          4. others (credit, gifts, etc.)

      C. Money in savings

 II. Between Subsystem Flows

      A. Human consumption

      B. Animal consumption

      C. Crop production

          1. inputs

          2. outputs

          3. quantities in storage

      D. Animal production

          1. inputs

          2. outputs

          3. quantities in storage

_____

62

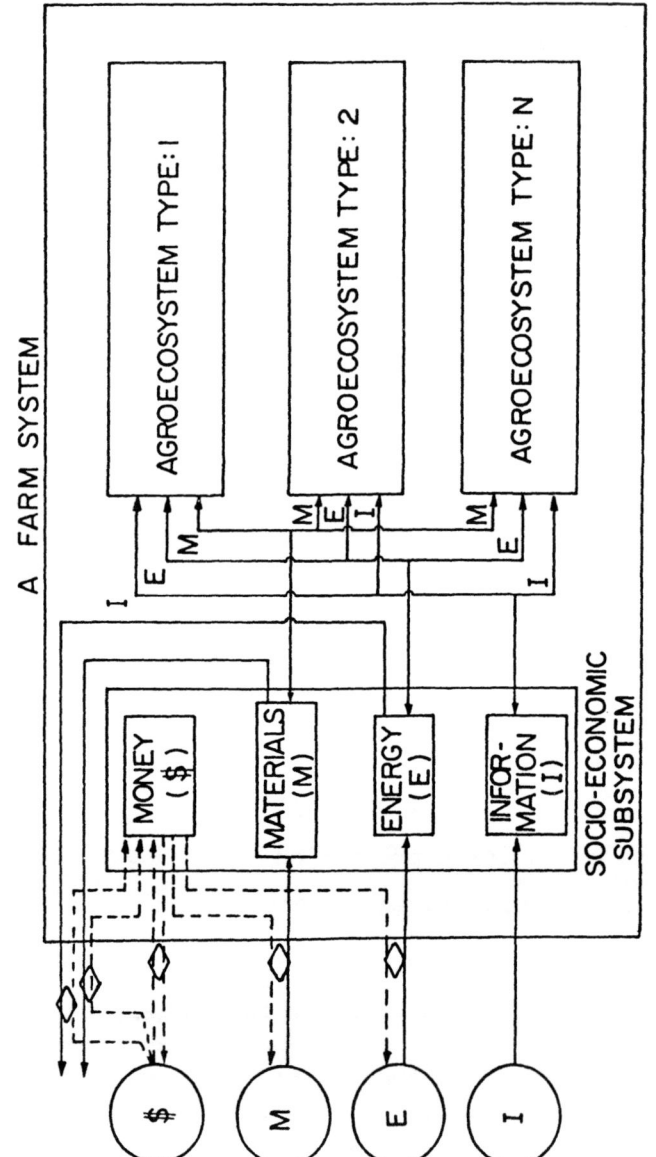

Fig. 1. A generalized qualitative model of a farm system with socio-economic and agroecosystem subsystems and inputs, outputs, and between-subsystem flows of money, materials, energy, and information.

inspected graphically.

The quantitative model and the dynamic flows were used to define a general farm management strategy used by Mr. Alvarado. Fifteen other farmers living at Yojoa were interviewed to determine if the farm system that had been analyzed was representative. Guidelines for the crop systems research in the Yojoa area were then developed.

Results

The quantitative model shown in Figure 2 shows a general overview of the farm system analyzed. Some inputs, such as food not produced on the farm and household articles, have been combined in order to reduce the complexity of the model.

Most of the farm system input and output flows were associated with the flow of money. A total of $1,830 (U. S. dollars) was earned by selling maize, rice, eggs, family labor, and by renting oxen and an ox cart. Total farm money input for the year, including $75 in credit, was $1,905. Total money output for the year was $1,648. Household articles (especially clothing) and food were a major expense (45 percent). Agricultural production-related inputs, including agricultural chemicals ($117 for fertilizer, $11 for herbicide, and $2 for insecticide), an ox cart ($200), and labor ($278) accounted for 55 percent of the money output.

The total inputs and outputs to the various farm agroecosystems are also summarized in Figure 2. The total labor (man-days), oxen energy (oxen-days), agricultural chemicals, seed, and crop production are in units/agroecosystem (as opposed to units/ha). In a few cases, such as labor inputs to the pasture plus oxen, chicken, and tree agroecosystems, data were not collected. This oversight was a result of not including these flows in the original qualitative model.

The farm system was characterized by strong interaction between the agroecosystems. In many cases the output from one agroecosystem was an input to another. For example, the pasture plus oxen system produced 181 oxen-days (OD) of energy. Of this total, 90 OD (50 percent) were used in the maize-maize sequence agroecosystem, 25 OD (14 percent) were used in the rice-bean rotation agroecosystem, and 66 OD (36 percent) were sold (oxen rented for plowing and hauling). The maize and rice consumed by the chickens were produced by the rice-bean and maize-maize agroecosystems.

It is difficult to analyze the agroecosystems in purely economic terms since many of the inputs are outputs from other agroecosystems and their real values (opportunity costs) are not shown. For example, if the maize and rice inputs to the chickens were worth the same per kilogram as the maize and rice sold in the market place and if the opportunity cost of the labor input is

64

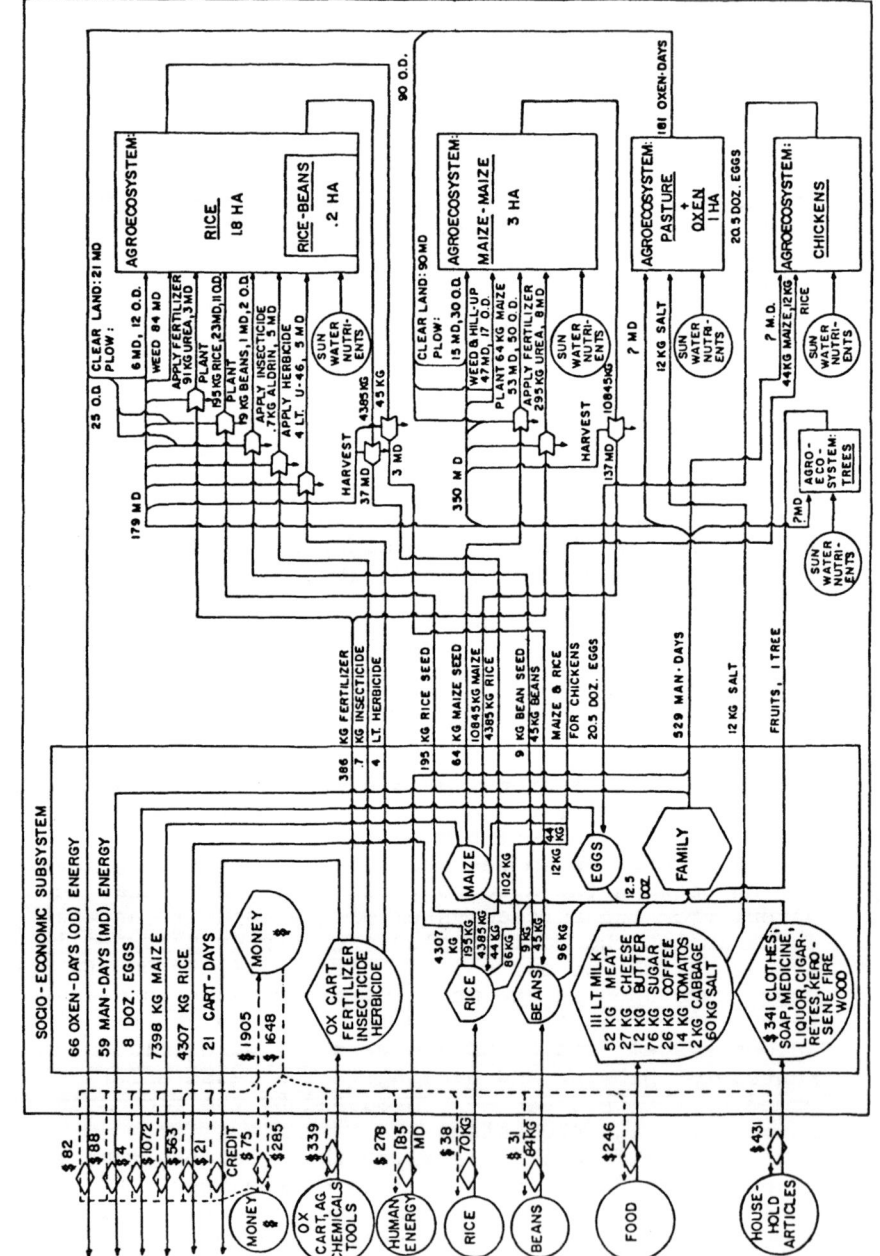

Fig. 2. A quantitative model of a farm system at Yojoa, Honduras with inputs, outputs, and between-subsystem flows shown as yearly totals. (Symbols after Odum, 1971.)

assumed to be zero (since children usually took care of the chickens), the inputs and outputs to the chicken system would be $8 and $10, respectively. However, if the maize and rice fed to the chickens were not of edible or marketable quality, as was often the case, the value of the inputs would be less. Also, the value of having chickens available to sell if an unexpected economic need occurs (risk aversion) is even more difficult to quantify.

Although the labor input to the pasture plus oxen agroecosystem was not quantified, the fact that young children of the family took care of the animals suggests that the opportunity cost of this labor was relatively low. The 12 kg/year of salt given to the oxen was worth only $1.50. Assuming a price of $1.33/OD, the 181 OD of output from the system was worth $240/ha. The maize-maize and rice-bean agroecosystems produced net returns of $287/ha and $115/ha, respectively (subtracting market value of the inputs from the market value of the outputs). One of the reasons for the lower return from the rice-bean system was that beans were only planted on 10 percent of the area planted in rice, while in the maize-maize system both maize crops were planted on 100 percent of the three hectares used for the agroecosystem.

While the quantitative model shown in Figure 2 gives an overview of the farm system, it does not show the dynamic chronological fluctuations of the farm system. Many flows had bimodal fluctuations. An inspection of the weekly data showed that money, labor, maize, and precipitation were probably the flows that most determined the general chronological fluctuations in the farm system.

Input, storage, and output of money for the farm system is shown in Figure 3. Two peak periods of money input to the farm system (gross income) occurred in October and in March. During the October peak there was a corresponding high output of money (farm expenses), but the output was less than the input, and farm savings increased. During the March peak, there was even less output, and savings increased even more. At the end of the study cash savings were much higher than at the beginning.

The bimodal money input fluctuations were due to the harvest and sale of maize and rice in September and October (first cropping period of the year) and the harvest of maize in March (second cropping period). The two cropping periods are undoubtedly a reflection of the rainfall pattern in the area (Fig. 4). The money input in March may have been higher than usual for that time of year because of the better-than-average maize production that occurred as a result of unusually high rainfall during January and February. The usual practice at Yojoa is to plant less maize and use less fertilizer during the second cropping period than during the first, since there is a high risk of drought during the second period. The

Fig. 3. Weekly input, output, and saving of money in a farm
system at Yojoa, Honduras over a one-year period.

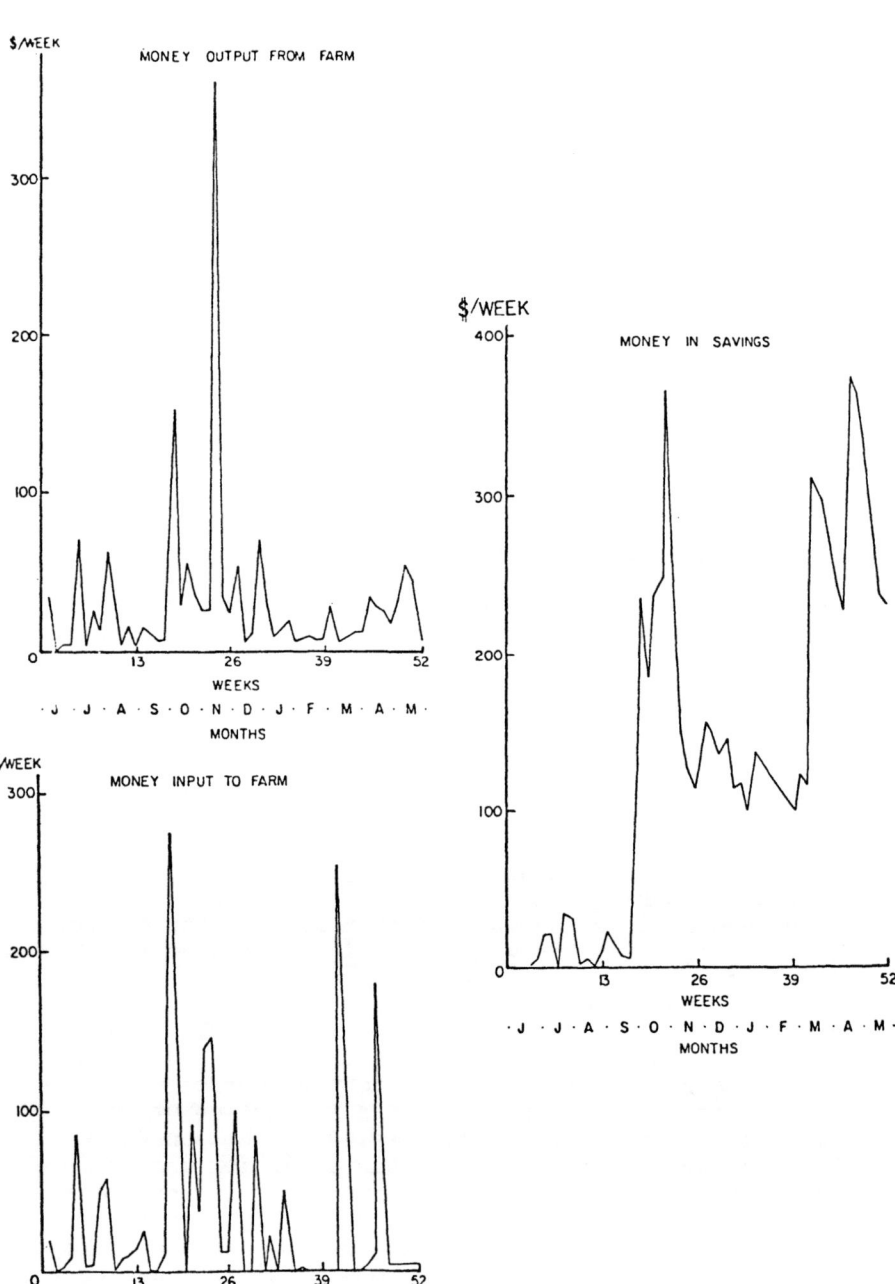

Fig. 4. Monthly precipitation at Yojoa, Honduras between June
         1976 and May 1977.

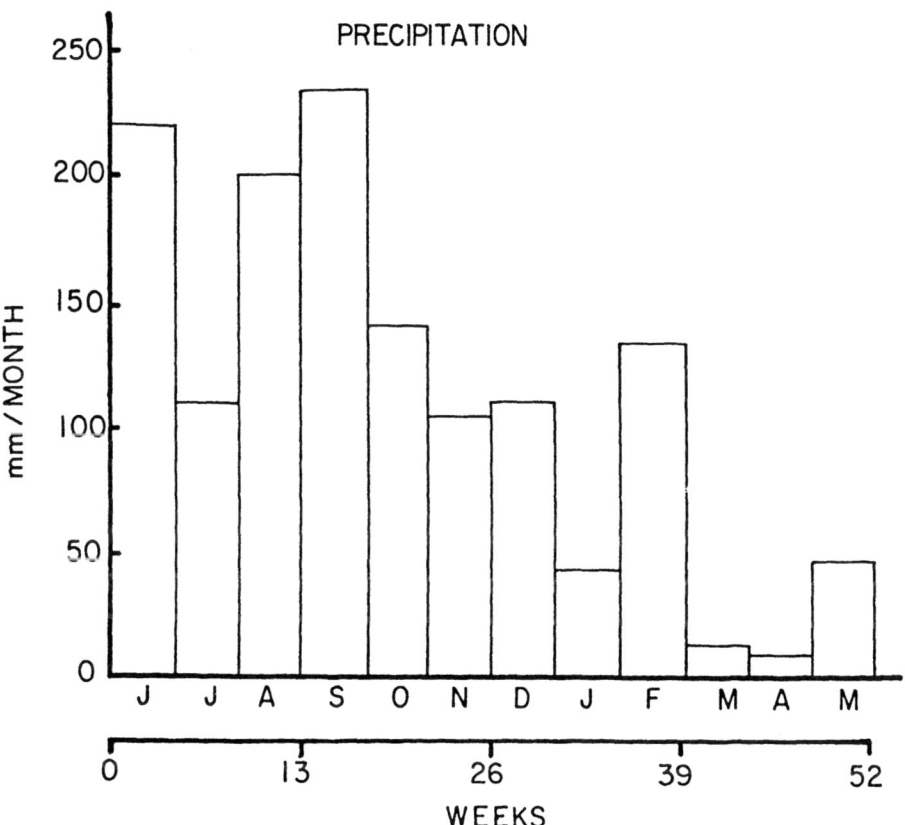

year before the study began many farmers at Yojoa, including Mr. Alvarado, lost their second maize crop. This may account for the difference in money in savings between the beginning and the end of the study.

The storage of large quantities of maize was an important aspect of the farmer's management strategy. When maize was harvested, approximately 50 percent was sold immediately and 50 percent was stored in the house. The farmer used his stored maize as a bank account, selling small quantities to meet household expenses (13 sales of less than 50 kg) and large quantities to meet larger farm management expenses (9 sales of 200 kg or more). Some of the stored maize was also eaten every day (3 kg/ day; 0.4 kg/day/person) and some was used as seed.

The fluctuations in stored maize over the one-year period can be observed in Figure 5. The rate at which the stored maize decreased was a reflection of economic and nutritional needs. The rate of decrease may also have been a reflection of the farmer's perception of the potential yield of his maize in the field. If environmental conditions were such that he could expect good yields (a high input to his storage area), the farmer would probably sell larger quantities and at a faster rate than if he expected low yields.

Figure 6 is a summary of the dynamic fluctuation in labor input and output and on-farm labor use. In general, more labor was hired during the first cropping period than during the second period because of the high amounts of labor needed to weed rice. Approximately equal amounts of labor were hired for rice and maize cultivation even though only two hectares were planted in rice and six hectares (3 hectares planted twice) were planted in maize. September, October, December, January, and April were the months with the lowest labor demand. As would be expected, labor need was the highest during the planting and harvesting periods.

Guidelines for Crop Systems Research

Before an attempt was made to use the results of the farm system study as a guideline for the crop systems research at Yojoa, the general farm management strategy used by Mr. Alvarado was compared to that of his neighbors. Because of the importance of maize in the farm system studied, Mr. Alvarado's strategy of storing large quantities of maize and planting, eating, and selling the maize in small quantities to meet household costs and in larger quantities to meet farm costs was used as an indicator of his farm management strategy. In a random sample of 15 farmers chosen from a group of approximately 40 farmers attending a field day, 60 percent had a strategy identical to Mr. Alvarado's. The other 40 percent differed only in quantity of maize sold to meet farm costs. This group only sold maize in large

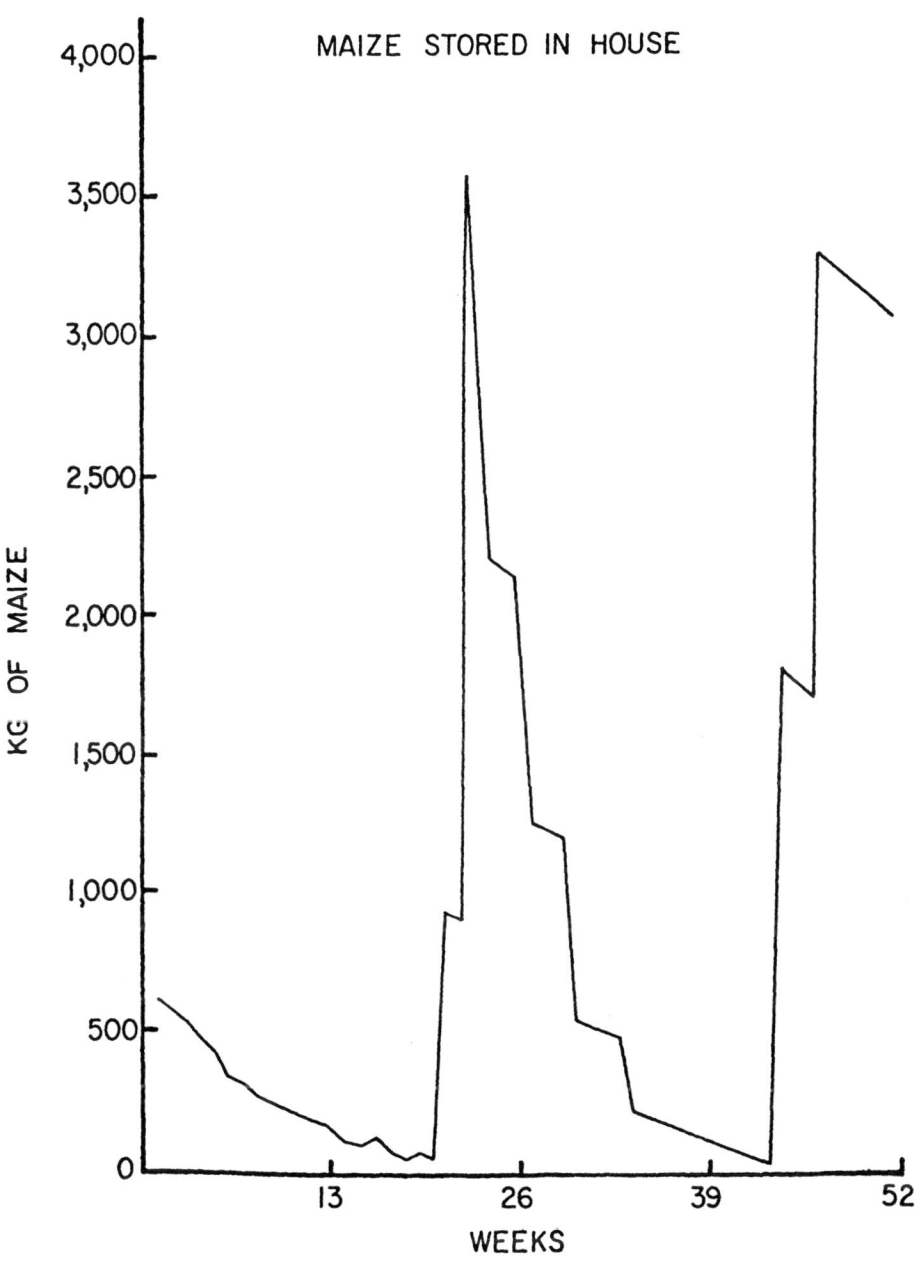

Fig. 5. Weekly quantities of maize maintained in storage in the socio-economic subsystem of the farm system.

Fig. 6.   Weekly labor input to the farm system and family labor
on and off the farm.

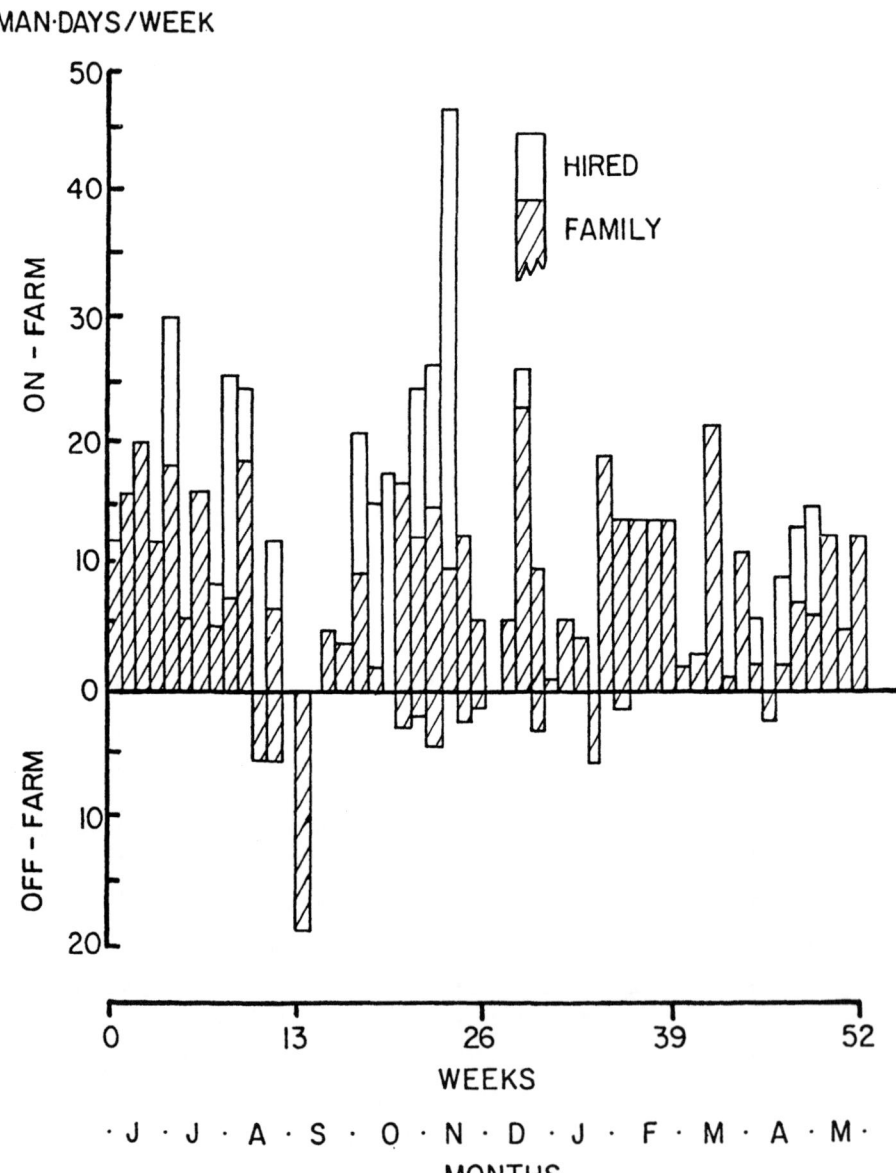

quantities to meet farm costs, and did not sell small quantities of maize to meet household costs. In no case was the price of maize in the market place stated as a reason for selling maize, even though during the year of the study, the price of maize fluctuated by more than 100 percent.

The following is a list of some of the general conclusions and guidelines resulting from the study:

1) Maize is an agronomic, economic, and socially important component of Yojoa farm systems and any changes suggested should not require the substitution of another crop for maize or a reduction in maize yield.

2) Maize, rice, and bean yields are highly variable and an effort should be made to design crop systems which could reduce the risk associated with existing crop systems.

3) Beans are not ecologically adapted to the Yojoa environment and other legumes should be tested to see if they could be substituted for common beans.

4) Weed control in rice is very labor demanding and herbicides should be tested as a way of decreasing labor need.

5) The existing crop systems use less labor in August, December, and April and alternative crop systems should be designed to take advantage of this labor surplus.

6) Few vegetables are produced or consumed in the area and crop systems with vegetable components or the design of household gardens should be considered.

7) No industrial or high-value cash crops are grown in the area and their potential should be studied.

The on-farm research of the crop systems project concentrated on finding alternatives to the maize-maize and rice-beans crop systems analyzed in the farm system study and to a maize and squash intercropped system that is common at Yojoa but was not part of the farm system study. After three years of research on spatial arrangements, varieties, and fertilizer modifications, the best alternatives generated were a) cowpea relayed between two maize crops planted in sequence; b) rice and maize intercropped followed by cowpea; and c) maize intercropped with pipian (a cucurbitaceae with high market value) planted twice in one year. The data collected in the farm system study were used to compare the potential of these alternatives with the system the farmers are presently using. These alternatives and the experiments conducted at Yojoa from 1976 to 1979 are described in CATIE mimeograph publications (1979a, 1979b, and 1979c).

Conceptual and Methodological Implications

An important objective of the farm system study conducted at Yojoa was to evaluate the general farm system concepts (Fig. 1) and the qualitative-to-quantitative model methodology. Given the total time dedicated to carrying out the study (one hour/week for 52 weeks), the quantity and quality of the data were very satisfactory.

As a data quality check, the money and maize that the farmer stored in his house was measured using two different estimates. Every week, the farmer was asked for his estimate of money in savings and of stored maize. These data were also estimated by adding inputs and subtracting outputs. At the end of the study the two estimates of money in savings differed by less than $150 (13 percent of the total money turnover). The maize estimates differed by 1300 kg (12 percent of the total maize turnover).

The questionnaire for this study was designed on the basis of a generalized qualitative farm system model and some preconceived ideas on the importance of certain components of the farm system. The study could have been improved by using a qualitative model of the specific farm system under study, rather than the generalized model, as a basis for the questionnaire. A farm-specific model could be formulated after a few preliminary visits to the farm.

After a number of farm system studies of this type have been done in a specific area, it should be possible to identify and separate static and dynamic flows. Estimates of the static flows could be made less frequently and this could reduce the interview time.

While farm systems are indeed complex, the conceptualization of a farm system as a set of subsystems with inputs, outputs, and between-subsystem flows that was used in this study was a valuable simplification tool. The formulation of qualitative and quantitative static models and the inspection of important dynamic flows was a successful methodology, and the usefulness of the data collected in this study demonstrates the potential of farm systems research.

Acknowledgements

The results reported in this paper were part of the Small Farmer Cropping Systems Project conducted by the Centro Agronómico Tropical de Investigación y Enseñanza (CATIE) and the Ministerio de Recursos Naturales (MRN) of the Government of Honduras and financed by the United States Agency for International Development, Regional Office for Central American Programs (USAID/ROCAP). José Nery Mayorga, an agronomist with MRN, conducted the last six months of the one-year interviews. The participation of the scientists of the Annual Crops Program at CATIE in the on-farm research phase of the study reported in this paper is gratefully acknowledged.

References

Centro Agronómico Tropical de Investigación y Enseñanza. 1979a. Descripción y evaluación del sistema de cultivos (maíz + arroz) - frijol de costa: una alternativa para el sistema arroz-frijol practicado por los agricultores de Yojoa, Honduras. CATIE, Turrialba, Costa Rica. 135p.
Centro Agronómico Tropical de Investigación y Enseñanza. 1979b. Descripción y evaluación del sistema de cultivos maíz/frijol de costa-maíz: una alternativa para el sistema maíz-maíz practicado por los agricultures de Yojoa, Honduras. CATIE, Turrialba, Costa Rica. 117p.
Centro Agronómico Tropical de Investigación y Enseñanza. 1979c. Descripción y evaluación del sistema de cultivos (maíz + pipián) - (maíz + pipián): una alternativa para el sistema (maíz + ayote) - (maíz + ayote) practicado por los agricultores de Yojoa, Honduras. CATIE, Turrialba, Costa Rica. 114p.
Odum, H. T. 1971. Environment, power, and society. Wiley Press, New York. 331p.

6
A Cropping Systems Research Methodology for Agricultural Development Projects

Hubert G. Zandstra

"Rural areas have labor, land, and at least some capital which, if mobilized, could reduce poverty and improve the quality of life. This implies fuller development of existing resources, including the construction of infrastructure such as roads and irrigation works, the introduction of new production technology, and the creation of new types of institutions and organizations" (World Bank, 1975).

Since the publication of this outstanding policy paper, the World Bank has encouraged rural development by helping to finance numerous area-based development projects. The same policy paper highlights the difficulty with which agricultural research results reach poor farmers and cites the common failure of researchers to treat small-scale farming as a system of cultivation that demands a comprehensive on-farm approach for technological improvements. An important reason for this is that traditionally research goals were generally formulated within disciplines. As the question is raised, however, of how the results of discipline-oriented research should affect food production and the efficiency of the farm enterprise, the relationship between research goals and the final recipient of technology, the farmer, becomes much less clearly defined.

The rate of technology change is increasing. New agricultural chemicals, new varieties and crop types with different tolerances for adverse conditions and a wide variety of vegetative periods, and new crop establishment and management alternatives are being developed in unprecedented quantities. The combination of these technological components into viable agricultural production methods is becoming increasingly difficult. For example, the replacement of a 150-day rice variety with one that matures in 105 days has tramatic effects on the production system of a farmer (Magbanua et al., 1976). Adjustments have to be made to nearly every farm operation.

As the simple replacing of one technological component with another has proven unsatisfactory, more of our agricultural research needs to be devoted to a careful synthesis of the new technology components so that crop production methods are efficiently adapted to the farm environment. The goal of agricultural research is, after all, to formulate improved production recommendations that are acceptable to farmers. To be acceptable, new production methods must satisfy a great number of requirements such as a good economic performance, a reasonable fit to farmers' resources, stability of performance over time, and a minimum of future research required for their maintenance.

My paper is about production technology and some of the methodological aspects associated with its generation. It presents a way in which the results of crop production research can be made more relevant to poor farmers, and pleads for the consideration of this or similar approaches in the planning and execution of agricultural development projects.

Technology-Environment Interactions

Crop production can be considered to be the result of two multidimensional vectors, the environment (E) and management (M), so that

$$Y = f(M, E) \tag{1}$$

Depending on the performance criteria, for example net gains, marginal returns to production factors, or returns to the farm enterprise, this relation can be transformed so that Y becomes a function of M, E, and costs. In formulating a recommendation, optimization processes are used to choose the input level of M. Obviously, the most appropriate input level will depend on the type of environment because of interactions between M and E in Equation 1. A simple example is that phosphorus fertilizer requirements for rice production are low on soils that are high in available phosphorus. A more consequential case is that double cropping rainfed lowland rice in regions with more than 200 mm rain for six months may be possible in heavy textured soils but not in light textured soils.

Recommended production methods must therefore be conditioned by the environment for which they are recommended. In effect, ignoring the technology-environment interactions increases costs of production and lowers returns derived from the recommendation. This in turn strongly increases the risks associated with the adoption of this technology. Without fine tuning new production methods to fit the physical and socio-economic environment of the farmer, probability of farmers' adoption will be severely reduced and the benefits derived from investment in agricultural research and extension will only be a fraction of their potential.

A lack of a well-defined methodology for farmer-level multiple cropping research has hampered the realization of effective on-farm research during the last decade. But a substantial number of researchers have recently contributed to the formulation of needed method-ology (Laird, 1968; Houser, 1970; Cady, 1974; Baker and Norman, 1975; Zandstra et al., 1975; Harwood, 1976). Many of these approaches have been applied in rural devel-opment projects such as the Puebla project and the Colombian rural development projects (Zandstra et al., 1979). The study of rice-based cropping systems at IRRI led to the formation of an Asian Cropping Systems Working Group, which has incorporated the results of these ex-periences in a cropping systems research methodology (Cropping Systems Working Group, 1975, 1976).

The cropping systems research methodology had to satisfy several requirements. First, the type of re-search had to be related to the production environment addressed. In this way a close fit of technology to physical and socio-economic limitations and opportunities could be achieved. Sufficient understanding of the envi-ronment would aid in extrapolation of results.

Second, farmers should participate in the design and testing of new multiple cropping technology. This would ensure early feedback from farmers about input, manage-ment, equipment, or market related constraints to the adoption of potential production alternatives.

Third, the research had to be multidiciplinary. The team had to combine capabilities in soil and crop sciences, crop protection, and agricultural economics.

Fourth, the methodology had to provide a clear iden-tification of the different tasks to be executed at the site. Hence, the responsibility of the different disci-plines among the research team members had to be recog-nized for each task.

The basic components of IRRI's cropping systems program are shown in Figure 1 and are described below.

Selection of Sites

The test sites should be carefully selected. They should represent major agroclimatic zones, so that results have a good chance of being applicable to other areas with the same environment.

An important criterion for site selection is the estimated potential for crop intensification. The est-mate is based on knowledge about the relationship between the environment and the crop intensification potential of several agroclimatic zones. Undoubtedly, the extent to which the potential for crop intensification can be esti-mated depends on how well this relationship is understood and how well the environment is defined. In effect, the estimate involves the same process as that described for cropping systems design, but it uses limited information about the environment. Continual interpretation of cropping systems research results obtained from different,

Fig. 1.   Components of IRRI's cropping systems program.

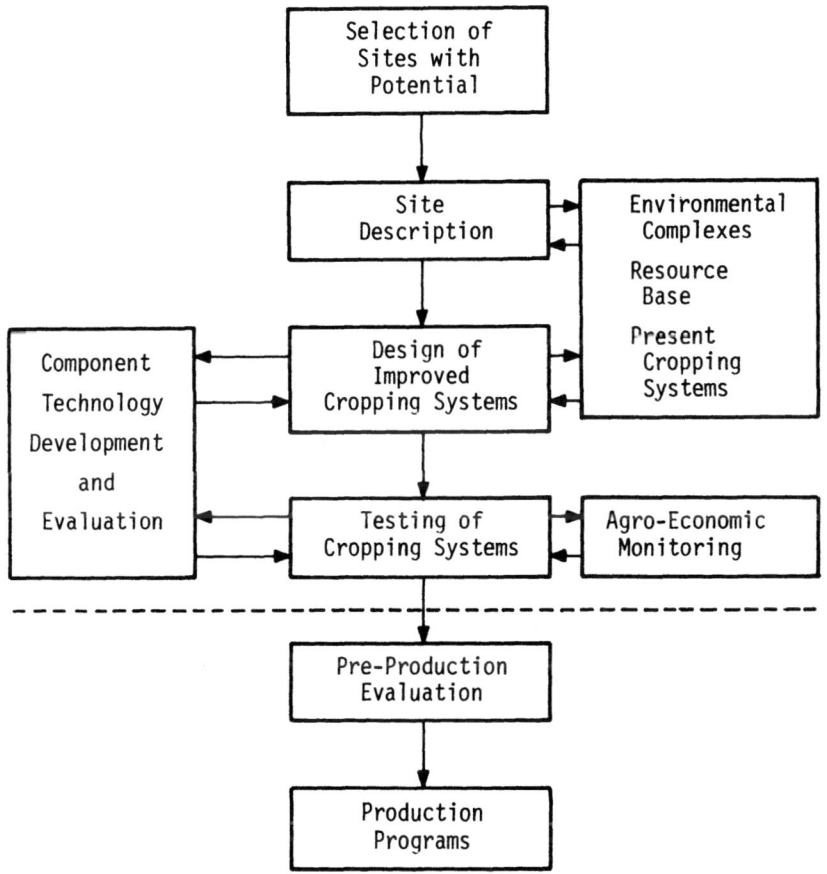

well-described (see next section of this paper) environments will provide the source material for a more precise classification of cropping systems potentials.

Site Description

The first activity of the cropping systems researcher is to describe the existing cropping systems in a selected area. The researcher needs to identify the different production complexes of the region and to relate them to physical and economic differences in the environment. An example of environment classification based on environmental complexes (the production complex was dominantly rice-fallow) is that used in the IRRI-BPI (Bureal of Plant Industry, Philippines) site at Iloilo. There, soil texture and landscape position were used to classify the environment.

A useful framework within which to relate these factors to cropping systems potentials follows (Zandstra, 1976).

First, environmental factors include physical resources (climate- and land-related), economic resources (availability of land, labor, cash, power, equipment, and materials) and socio-economic conditions (product prices, input costs, marketing costs, and customs reflecting preferences for certain foods or management practices).

Second, the cropping systems researcher specifies the factors he or she wants to operate on and those to consider invariant. The first set will be included in the management vector (subject to optimization), and the second set will be part of the environment vector of Equation 1.

Third, in environmental classification, readily modifiable physical factors should be excluded: nitrogen and phosphorus fertility; easily corrected microelement deficiencies; and the normal incidence of pests. The relation of $Y = (M, E)$ is thus reduced to one in which standard crop-management practices in M are assumed to correct for variations in the readily modifiable factors in E. Those factors remaining in E are cropping pattern determinants and should be used for environmental classification.

Fourth, a union of sites that have similar cropping pattern determinants is defined as an environmental complex or land type; a union of sites in which the relative performance of cropping patterns is substantially the same is defined as a production complex (Zandstra, 1976). A production complex is measured by cropping pattern performance and is, as such, an ecological unit. If the performance of cropping patterns is substantially different for any subset of sites within an environmental complex, one or more important determinants must have been overlooked in the description and specification of that complex. This provides the ability to test the

adequacy of the environmental description method employed.
Substantial progress has been made in the identification of physical cropping pattern determinants (FAO, 1971; IRRI, 1974), but their measurement and the measurement of associated pattern performance have been sadly lacking. In addition, the analysis and interpretation of research results have more often than not been related to the site and not to the environmental characteristics of the site.

The description and classification of the environment requires a contribution from land and soil classification specialists at an early stage of site research. The quality of the land, climate, and soil classification will determine the usefulness of the research results obtained beyond the direct project area.

Beyond the description of land type, site description includes a short baseline survey that describes crops, cropping patterns, and cropping systems and their association to land types. It also provides a summary of major farm types in the area, their holdings, labor and power resources, access to credit and agricultural chemicals, and their technological history. The baseline survey also evaluates wage rate variation throughout the year and the production methods and their results for a few major crops in the area.

Cropping Systems Design

In terms of Equation 1, cropping systems design is the specification of the management vestor M. The Asian Cropping Systems Working Group (1976) defined it as a synthetic activity that employs the physical and socio-economic site characteristics obtained at the descriptive stage, together with knowledge of the effect of those characteristics on the performance of cropping patterns, in order to identify intensified patterns that are well adapted to the site.

The design activity (Fig. 2) is focused on a certain land type. A limited assembly of practices from the available component technology can be employed in design. The technology includes cultivars; tillage practices; planting methods; plant population considerations; knowledge of optimal spatial relations between intercrops; crop interactions; effects of crop combinations and cropping sequence on weeds, insects, and diseases; water management methods; and pest control methods (by hand, pesticides, crop resistance, or escape). The technology also includes accumulated knowledge about the performance of cultivars and about the management practices listed above, under the conditions specified in the environment vector. Among those conditions are amount and distribution of rainfall and irrigation; landscape hydrology; drought, saturated soil, high precipitation and humidity during the crop establishment and harvest periods;

Fig. 2. Schematic presentation of the design of alternative
cropping systems for a selected environment.

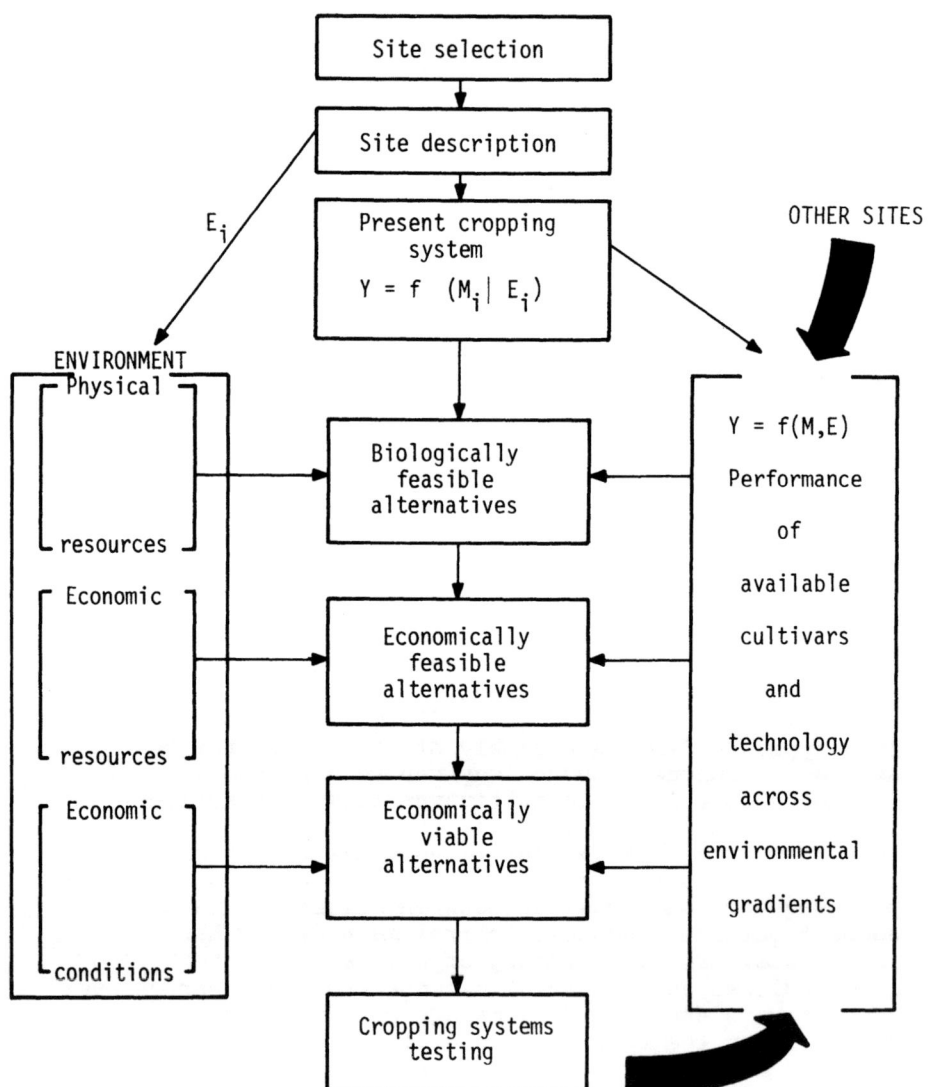

temperature and day length variations; extreme soil conditions; and predictable flooding.

The process of cropping systems design (Fig. 2) by
necessity employs certain performance criteria. Those
criteria should include estimates of cropping pattern
performance, the available resources, and a pattern's
resource requirements. A difficulty arises in determining the resources available to the cropping pattern. The
resources are most easily determined by substitution;
slack resources of the farming system are added to the
resources used by the cropping pattern that is to be
changed.

Design of the Site-Related Research Program

The formulation of the research program for a site
coincides with the design of cropping patterns for that
site and should be completed at least one month in ad-
vance of the first seeding date at the site. Normally,
the yearly research program is discussed at a workshop in
which all researchers at the site participate. Site
researchers should be given prime responsibility for the
presentation of previous research results, and should be
encouraged to contribute their insights on the existing
farming systems, the potential for increased production,
and farmers' reactions to alternatives. The workshop
should draw on the support of senior cropping systems
scientists and subject matter specialists in some or all
of the areas of economics, entomology, weed science,
water management, plant pathology, soil fertility, and
plant breeding. This workshop may take about three days
and although the research program for the site is designed before the cropping season starts, it may be useful to
re-evaluate the research program after each crop and make
the necessary modifications.

Cropping Pattern Trials

Four steps are suggested for the design of the cropping patterns to be tested at the site.

First, decide upon the land types to be studied at
the site and describe each of these as precisely as possible. The team need not conduct research on all land
types in their area of operation; generally by using two
to four of the most important (common) land types, the
team can cover the vast majority of production systems
at the site.

Second, identify variables that constrain crop production, such as fertility problems, minor element deficiencies or toxicities, or the common occurrence of crop
pests.

Third, decide upon the cropping patterns to be
studied for each land type. These patterns should be
carefully designed in accordance with the physical and

socio-economic conditions prevailing at the site. The farmer's cropping history, climate, product value, and potential market are all important factors to be considered.[1] For each land type the research team should limit itself to three or four cropping patterns. These patterns may be the same for different land types. In fact, it is desirable that the performance of one or more patterns can be compared between land types.

Fourth, each cropping pattern needs to be assigned a management technology. Figure 3 is an example of the complexity of a cropping pattern and the information required with respect to component technology. As the research team considers different alternatives, it must evaluate the expected response and the cost involved for each alternative. After the design of the cropping pattern, a simple cost-and-return analysis must be conducted. These factors should not be taken lightly, as it has been estimated that to decide upon varieties, pest management, fertilizer additions, and methods for tillage, planting, weed control, and harvest, in addition to the timing of all operations, more than 30 decisions need to be made for a two-crop cropping pattern.

The input levels for component technology assigned to the cropping pattern should be such that they will increase net returns above those obtained from existing patterns and still provide returns to purchased inputs and labor that are above those normally obtained in the region.[2]

During the first year, the component technology chosen for the cropping patterns will depend primarily on information from the environmental description and previous research at the site and in similar sites. In time more information on component technology will become available from research at the site and will increasingly form the basis for decision making about the component technology levels to be used for the cropping patterns. Example specifications for weed control component technology for a site are presented in Table 1.

---

[1] See information required to design and test for economic criteria, page 36a to 36c, *Fourth Cropping Systems Working Group Report, 1976*.

[2] Large-scale credit programs for crop production can substantially reduce the cost of production capital in a region and the returns farmers demand from purchased inputs. Although the extent of such changes are hard to predict, where such credit programs are foreseen, returns to purchased inputs may be somewhat below those obtained in the present production system in the absence of a credit program. They should, however, always be above the real cost of credit.

83

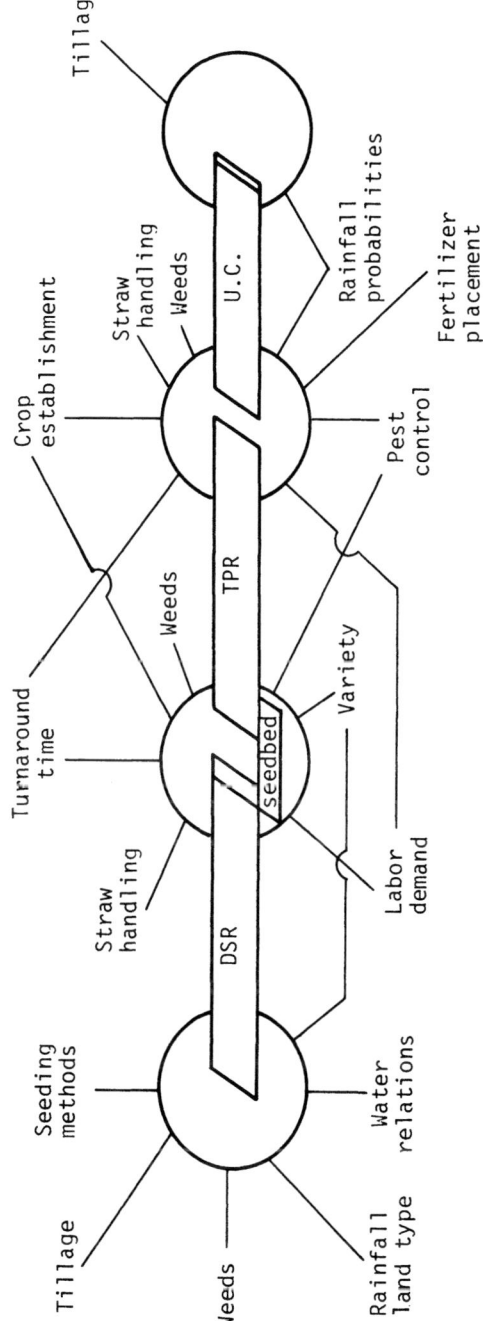

Fig. 3. To assign component technology to a pattern requires a careful selection from many alternatives. DSR = dry seeded rice, TPR = transplated rice, UC = upland crops.

Table 1. Recommended weed control practices for cropping patterns, Pangasinan, 1977-78.

Crop	Weed control methods	Rate (kg a.i./ha)	Time of application
Corn (before rice)	Hilling-up, 2 passes	–	3 WAE* or just after fertilizer topdressing
Dry-seeded rice	Butachlor followed by one hand-weeding	2.0	Immediately if soil is moist, or if soil is dry, after germinating rain followed by "as needed"†
Wet-seeded rice	Well puddled seedbed. If there is standing water - no weeding; otherwise, spot weeding	–	As needed
Transplanted rice	Well puddled seedbed. If there is standing water - no weeding; otherwise, spot weeding	–	As needed
Upland crop			
Field not plowed	Paraquat to be applied if 50% plant cover at time of crop establishment; otherwise, no weed control	0.75	Prior to furrowing
Field plowed	Mungbeans and cowpeas - no weeding	–	
	Sorghum - interrow cultivation	–	To 4 WAE

*WAE - weeks after emergence

†Refer to manual weeding or spotweeding as needed.

Cropping Systems Testing

Cropping patterns and their management are tested in farmers' fields to verify the assumptions made in the cropping systems research process, particularly those at the design stage. The assumptions are:

1) The proposed cropping system is biologically suited to an important physical environmental complex of the site. Yields of crops in the pattern should therefore be adequate, and biological instability should not occur.

2) The cropping pattern's requirements for economic resources, such as cash, labor, and power can be met.

3) The management components of the specified patterns are economically optimal.

4) The cropping patterns satisfy the selected economic performance criteria.

Performance Criteria

The first step in the testing process is to define satisfactory performance criteria (Fig. 4). To be useful in the context of site related research, these should not require complex computations. Nonetheless, the performance criteria must be conditioned by the factor costs prevalent at the site and the present knowledge of farmers' decision making. Because of farmers' control over on-farm resources (land, farmer's time, family labor including exchange labor, water, and farm implements), the net returns to these resources form a useful first estimate of the overall benefit derived from a cropping system by the farm enterprise. Further performance evaluation can be based on returns to cash and labor compared to their cost in the region; cash requirement compared to its availability; the required level of indebtedness compared to actual cash income of the farm; and risk as a function of yield variations (preferably the subjective estimates of farmers) and levels of cash input (Zandstra et al., 1975).[3]

The testing process requires more time and research personnel than the other activities described in the

---

[3] Recent work on opportunity cost budgeting methods (Price and Barker, 1977) has led to a relatively simple method for handling seasonal variations in labor wage rates. In-depth studies in whole farm budgeting techniques are being used to find ways in which we can condition simple partial budgeting techniques, or their interpolation, to farm types with different resource endowments.

Fig. 4. Testing of cropping patterns.

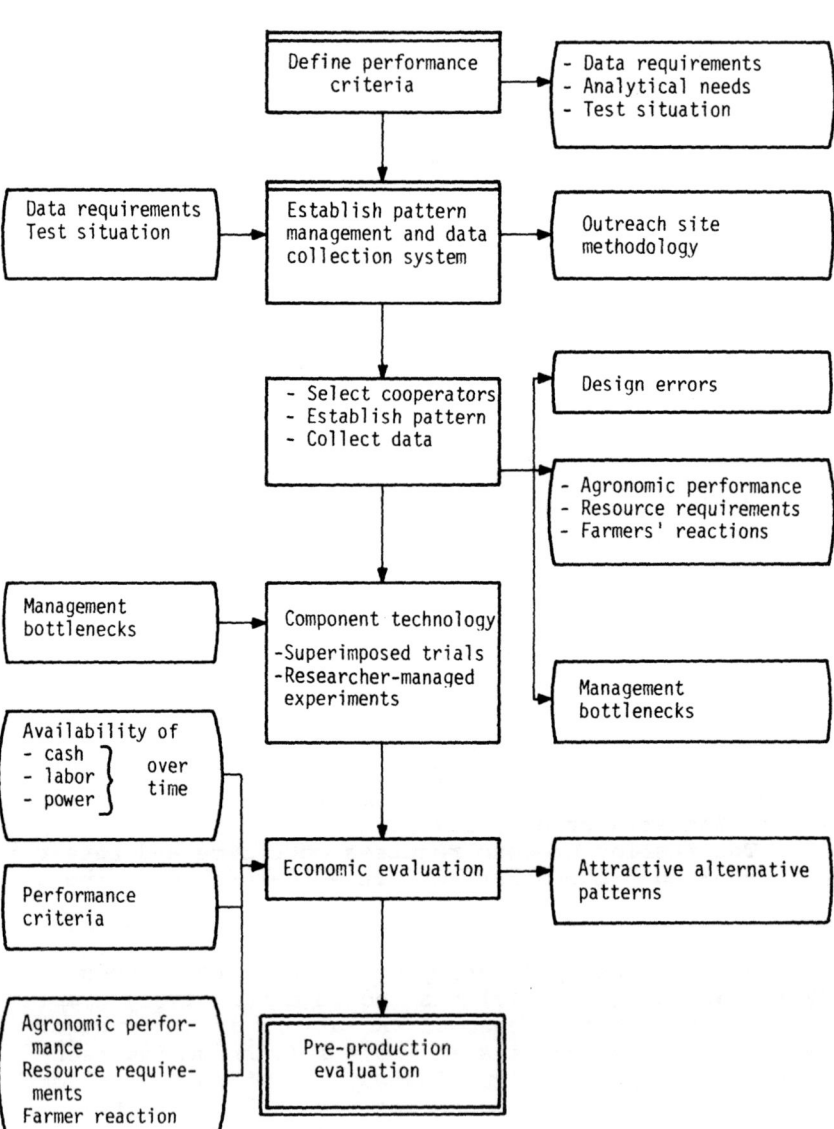

cropping systems research process (Fig. 1). The monitoring of patterns and the data collection system must be both manageable and sufficiently rigorous to allow reliable estimates of cropping pattern performance, its resource requirements, and the farmers' reactions to it.

Experimental Design

The trials compare patterns that differ in crop types, the number of crops, their establishment method, and time as well as their management. This makes it impossible to test patterns using replicated small plot experimental designs, as the objective is to evaluate cropping patterns on the basis of their performance in the land types for which they were designed; the land types become the experimental area and fields within the land types become the plots. The experimental design used is a completely randomized design in which replicates are assumed to sample the variation of field conditions existing within the land type.

These trials often involve new crops and a change in time of operation from that used in the existing patterns in the area. For this reason, the trials should be managed by farmers to evaluate the farmers' capability to manage the cropping pattern. This gives opportunities for the identification of conflicts between the operations required for the pattern and the farmers' resource base or the climate or land qualities. Cropping patterns are tested in large (1,000 sq. m.) plots to allow measurement of labor and time required for the operations used in execution of the patterns. This in turn allows precise cost-and-return analysis for the patterns.

For the design of cropping pattern trials, the following general guidelines are suggested:

1) The research team should select two or three land types on which to focus its research.

2) For each land type the team should select three cropping patterns to be evaluated. For some patterns on some land types, these patterns may be the same.

3) Each cropping pattern should be replicated in at least five fields in total and in at least four fields per land type.

The above research design should be modified as the team acquires more experience at the site. During the first year the number of patterns to be studied may be higher than three per land type. During the second year the number of patterns can be reduced and the number of replications can be increased to at least five in total and at least four per land type. During the third year the team should have focused in on the most promising cropping patterns. This will allow them to increase further the number of replications per pattern to at least six in total and at least four per land type (Table 2). It is recommended that the research team manage from

Table 2. Year to year variation in the design of cropping pattern trials reflecting trend towards reduced number of patterns and increased number of replications.*

Land type	Pattern								Total
	1	2	3	4	5	6	7	8	
Year 1									
1			4	5	4	5			18
2	4	5	4					4	17
3	4		4		4		4		16
Total	8	5	12	5	8	5	8	0	51
Year 2									
1			4	6				5	15
2	6	5	4						15
3		5	4					5	14
Total	6	10	12	6	0	0	0	10	44
Year 3									
1			4	6				4	14
2		6	4						10
3		6	4					4	14
Total	0	12	12	6	0	0	0	8	38

*The numbers in the tables are the replications (fields) of a pattern in a land type. For example, in Year 1 pattern 6 is replicated 5 times in land type 1.

40 to 50 cropping pattern trials.

Data Collection

The performances of experimental cropping patterns are compared to those of farmers' existing patterns, as the latter provide the research team with a measure of the cost and productivity of production factors in the area. Methods have been developed for the collection of climate, plot, crop, and operational records for experimental and farmers' cropping patterns. These records include time required for the operations and equipment or materials used. Where appropriate, specific variables such as depth of water or moisture condition of the soil can be monitored.

The testing phase allows evaluation of the research team's ability to design improved cropping patterns on the basis of the environmental classification employed. It allows an evaluation of the efficiency of the cropping pattern determinants as stratifying variables for design and future recommendations. In this manner the test results can lead to modifications in site description. In addition, the testing of cropping patterns on the farm provides important clues to techological constraints to increased production. These might include lengthy turn-around times between crops, a lack of techniques for upland crop establishment in previously puddled rice fields, weed control in dry seeded rice, fertilization of zero-tillage-planted upland crops growing on residual moisture, and ratooning rice varieties and management of the ratoon crop (IRRI, 1976; Zandstra and Price, 1977).

Component Technology Research

Although the major activity at a cropping systems site is the testing of improved cropping patterns, the site team must also ensure that the management specified for each of the crops in the patterns is optimal.

As the team discusses the component technology to be assigned to cropping patterns, it will also identify subjects on which there is a lack of information that needs to be studied at the site. This may be a need for further environmental description, such as better definition of the duration of irrigation, the time and frequency of rains, labor wage rates during harvest time, or the farmer's ability to identify insect pests. It often involves the need for better component technology such as varietal screening, insect, weed or disease control, fertilization, tillage methods, or the date of establishment of different crops. During the first year it is often useful to do time-of planting trials for the important crops at the site over their potential range of planting dates. These trials should be monitored for the occurrence of insects and diseases. An early definition of response to major plant nutrients is also required.

Component technology research is conditioned to the cropping pattern selected. It normally addresses only one crop of the pattern sequence and one or two variables, such as variety trials, tillage methods and subsequent levels of weed control, or method and rate of nitrogen application. Component technology trials are generally managed by the cropping systems researchers rather than the farmers.

The research team must be careful to study only those management components that have a major impact on the economic performance of the cropping pattern. Generally, the research focuses on the responses to inputs and leaves explanation of underlying mechanisms to the other physical and biological researchers.

Selection of Factors and Treatment Levels

For the initial experiments, three general sources of information should be used to identify factors and treatment levels to be tested: baseline surveys, a priori knowledge of crop requirements, and previous conventional field experiments conducted in the site area or in similar environments elsewhere. The latter may have been conducted in anticipation of a cropping pattern research program to follow or through the routine activities of organizations conducting multilocation trials. It is also advisable to identify the two management components that demand the most cash and the two components that require the most labor. Next, estimate the effect on yield of changes in each of these components, and evaluate the potential input savings or yield increases that could be derived from research on these factors.

Superimposed Trials for Component Technology Evaluation

Most component technology research should be closely associated with the cropping pattern tests and should be designed to test the present management assigned to the pattern. To ensure close association with the cropping pattern trials, much of this research should be conducted in the same fields in which the patterns are tested (hence, the term superimposed).

At present it is recommended that the designs for the superimposed trials satisfy certain objectives. They should: evaluate the return farmers derive from purchased material inputs used for weed control, fertilization, and pest and disease control; evaluate the return the cropping pattern component technology obtains from these inputs; determine whether possibilities exist for modification of the management components assigned to the cropping pattern for weed control, insect and disease control, and fertilization that lead to increased yield; and determine whether these yield increases are

sufficient to pay for the additional costs of the modified management components. To achieve these objectives,
superimposed trials must include the following component
technology levels: a simulation of farmers' management
level; farmers' management level without any purchased
material inputs; the level of component technology assigned to the cropping pattern; and a level of component technology that will produce higher yields than the cropping
pattern or that will produce similar yields at substantially lower input levels.

Various treatment designs can be used for superimposed trials, depending on the factors considered to be
of importance. These trials evaluate the performance of
the component technology across the land type and are
therefore normally not replicated within a field. Each
trial is established in five to eight cropping pattern
fields.

Researcher-Managed Trials

These trials are entirely managed by the cropping
systems research team. They evaluate in detail specific
management components to be assigned to cropping patterns.
They cover a wider range of management alternatives than
the superimposed trials. Thus, an increased number of
variables and levels are included in the treatments.
Researcher-managed trials seek to understand more precisely the type of responses to input levels and evaluate
high risk treatments about which too little information
is available to be included in cropping patterns managed
by farmers. The results of researcher-managed trials are
analyzed with an emphasis on treatment differences and
require considerable precision. These results determine
future changes in cropping pattern management levels and
the management components to be studied in the superimposed trials.

The experimental designs for researcher-managed
trials will not be discussed in detail. They follow the
considerations of small plot experimental design on
research stations. Because of limited field size, treatment numbers should normally be kept between six and
twelve. The number of replications should be three or
more, except where multilocation testing is involved, in
which case within-field replications should be reduced
to two, as long as the total number of replications is
four or more.

Researcher-managed trials can be conducted at research stations if the environment (climate, soils) at
the station is the same as that of the land type studied
at the site, or if the purpose is strictly to compare
treatment differences and no strong interaction with the
environment is expected. In such cases, the site research team requiring the information should encourage
researchers on the stations to conduct such experiments.

Whether conducted at a research station or at the site, these trials should use the same tillage methods and implements and the same component technology (for fixed management) as that used for the corresponding crop in the cropping trials. For factors that are varied, the treatment levels must include those used in cropping trials and the high level treatment of the superimposed trials.

Limits to seeding dates that apply to that crop in the cropping pattern must be applied to the component technology trials. This is important, as it will allow linking of the component technology research results to those of the cropping pattern trials. Where field x treatment interactions are considered important, the number of fields should be at least four and within-field replication can be reduced to a minimum.

Applied Research and Preproduction Testing

Applied research evaluates alternative cropping patterns at many sites that are representative of the environmental complexes for which the patterns were designed. The specification of the environmental complex is important. Applied research testing not only must provide extension or production agencies with alternative cropping systems with clearly specified management, it must also clearly delineate the situations to which those cropping systems are adapted. The domains of adaptation of recommended cropping systems must therefore be specified in terms that can be used to differentiate the action of production programs for different environments. That requires that the domain be mapped or associated with existing geographical boundaries or be described in site-differentiating terms, such as soil texture or drainage characteristics, that can be handled by extension workers on the basis of simple observation.

Preproduction testing follows applied research. It focuses on training of extension workers and on discovering the availability of credit, seed, and agricultural chemicals. In general, it identifies and prepares the institutions and personnel required for implementation of recommended practices on a wide scale. Preproduction testing also evaluates the performance of a recommended practice on a large scale.

One difficulty with production programs that seek to change farmers' cropping systems lies in the great variety of crops involved. Each crop has its own specific management package, its own credit and input requirements, and its own critical location in a cropping sequence and in a specific environment. That is a lot of information to carry through a delivery system, and the production program methods to be used will undoubtedly require critical assessment (Gomez, 1977).

Institutional Requirements of Site Related Cropping
Systems Research

At this time, the site related research method is
being applied by nearly 40 research teams throughout South
and Southeast Asia (Carangal, 1977) (Fig. 5). Many of
those teams receive advice and backup from regional or
central research station and university-based senior
staff in national programs. As the on-site research
proceeds, the capabilities required for the research
model become clear for all levels.

At the Site

The research team at the site is the instrument of
cropping systems research. It is the contact point between the research structure and the on-farm reality it
must address. The site team must therefore be able to
identify different environmental complexes based on land
types, textural differences, irrigation, drainage characteristics, and slope of the fields.
The team must be trained in farm survey methods to
determine the farm resource base and to identify the
existing management practices and their relation to important environmental factors at the site. It must relate to the farmers and be trained in the interpretation
of farmers' comments. In addition, the site team must be
able to plan and execute experiments, analyze them, and
interpret results. The site team also has to be involved
in the decisions made about the focus of its research.
For these reasons, it needs to participate in the definition of research priorities for the site and in the
planning of the experiments and surveys. It must be encouraged to become a strong multidisciplinary unit that
formulates hypotheses about the type of production technology required for the land types in the site--hypo-
theses that are continually tested against daily observa-
tions. The site-team should be a dependable source of
information about farm-level production techniques and
the performance of technical innovations in the area
covered by the site. It is particularly important that
the site team consult with local extension and irrigation
personnel, who can provide guidance in the selection of
cooperating farmers and provide details about the technological history of the site that are valuable to cropping systems researchers. Extension organizations should
also be exposed to research plans and on-farm trials at
an early stage.
The Cropping Systems Training Program at IRRI
carries groups of graduates from various disciplines
through the physical, biological, and socio-economic
aspects of site description, design, testing and component technology research, preproduction testing, and
production program formulation. The training employs

94

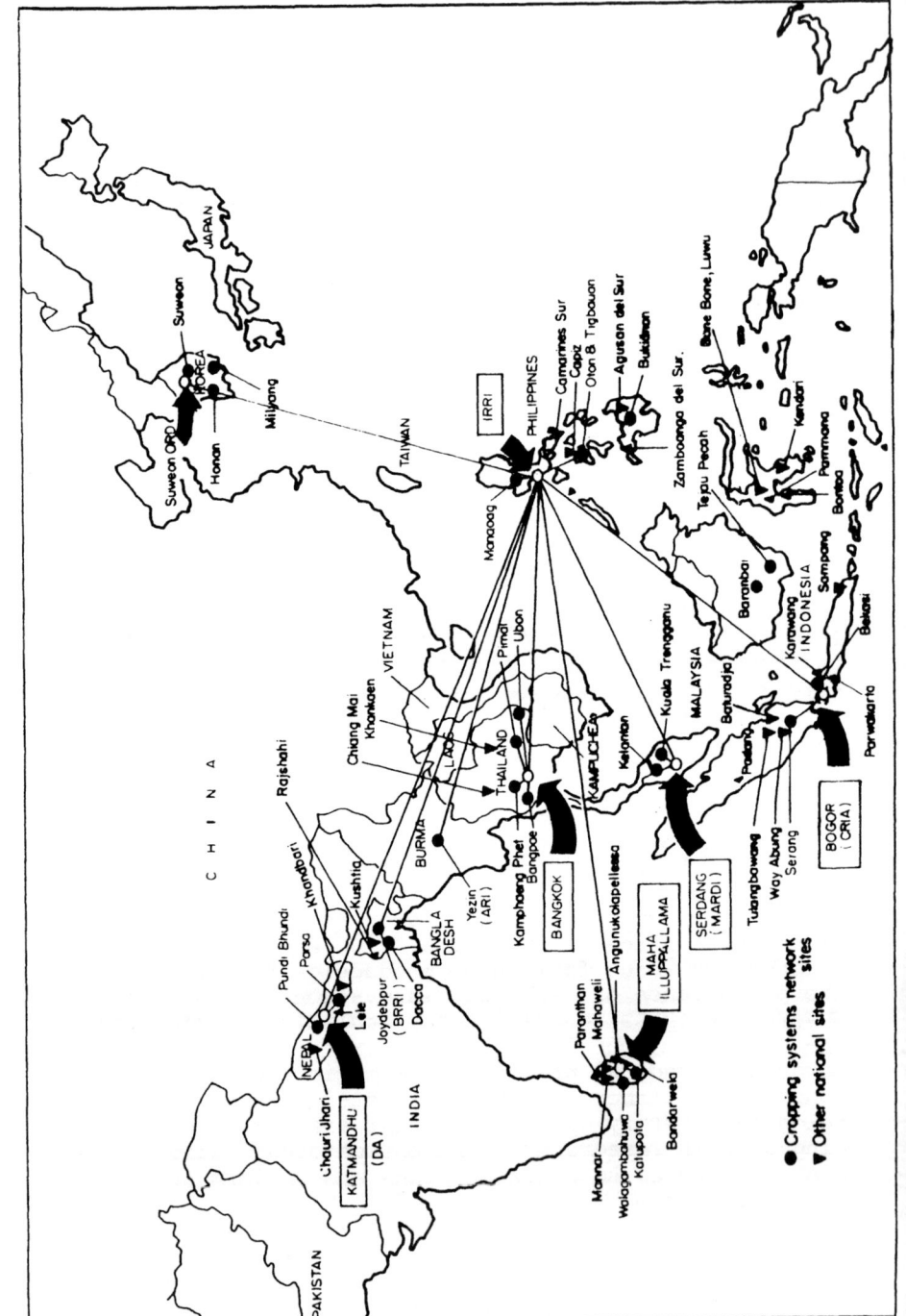

Fig. 5. Asian Cropping System Network.

examples and methods used at existing research sites and exposes trainees to several sites.

Regional and National Level Support

To operate the on-farm research at the site with the bachelor of science and the occasional master of science level staff, the team needs to be continually supported and encouraged. Our experience is that the teams derive strong motivation from the realization that they are addressing the real, everyday problems of farmers and that their solutions are immediately affecting the farmer-recipient group with whom they can identify. In addition to this motivation, the teams need to maintain contact with research institutions and recent research. They also need guidelines for environmental descriptions, research design, farm surveys, and experimental designs.

This requires a group of specialists at the research centers with experience in site-related research, in addition to the advanced training needed to advise research teams at the sites. These groups can often be composed of researchers working at existing regional or national experiment stations. Multidisciplinary team discussions at these stations can be encouraged and then such groups can work with a number of site teams offering support in research design, analyses, and interpretation. In addition to providing methodological and motivational backups to teams, the support group provides contacts with experts for consultations on specific problems, such as the identification of rare pests, minor element deficiencies, or disease problems.

Up to this point, cropping systems research has been discussed in terms of operations research designed to incorporate available knowledge, processes, and materials (biological, physical, human, and institutional) into crop production methods suitable for identified environments with clearly defined farm resource availabilities and institutional support structures. Because of the operational nature of site-related research, the project depends completely on technology available to it. This comes from national level experiment station and university research on one hand, and from the farmers in the region on the other hand. At the national level, there is a need for continued backup by commodity- and discipline-oriented researchers to resolve bottlenecks to increased production identified at the farm level (Fig. 4). In addition, the national institutes need to continue the development of on-farm research methods that will improve on-site operations in environmental classification, in research on soil and crop management and plant protection methods, and in the economic evaluation of production alternatives. To achieve this, commodity- and discipline-oriented researchers should visit on-farm research sites and invite opinions about research needs and priorities.

Results of research on rice-based cropping systems in the Philippines and other Asian countries have identified a shortage of information on:

1) Use of crop intensification techniques.

2) Crop establishment methods, particularly for upland crops after lowland rice.

3) Tillage methods, including the use of alternative tillage implements.

4) Interactions between land types and performance of cropping patterns.

5) Methods required to more effectively incorporate farmers into the on-farm research process.

6) Weed control techniques.

7) Effective methods to evaluate insect and disease occurrences and to condition insecticide recommendations to these.

8) Methods for identifying biologically stable cropping patterns.

9) Baseline survey methods to identify farmers' production techniques.

10) Methods to evaluate the performance of cropping patterns.

11) Methods for judging the institutional intervention required for the introduction of new multiple cropping technology.

12) Adequate description of the climate to allow crop scheduling.

Institutional Constraints to Cropping Systems Research

A new production technique is often constrained by institutional characteristics, because they were not designed to handle it. In the same way, the change from strictly discipline- and commodity-oriented on-station research to interdisciplinary multiple cropping-oriented research on farmers' fields is constrained by the existence of research institutions and traditions that were not designed to cope with the requirements for multiple cropping research.

The strong multidisciplinary nature of the site research teams requires the participation of agronomists, soil scientists, economists, and plant protection specialists. A similar, or still broader, multidisciplinary requirement exists for advisory support at the regional or national level.

In most countries, the capabilities in soil and land research, soil fertility and crop improvement, farm management economics, climatic analysis, and irrigation and water management are found in different institutions or agencies within the department of agriculture. This has made the structuring of the national programs based on multiple cropping research in the farm environment a difficult task. It requires that institutions responsible for the generation of new production technologies--not a variety or fertilization rate, but a completely specified and carefully tested sequence of crop and management activities--acquire capabilities in disciplines not normally represented among their staff. In addition, it requires considerable training and management planning to provide the operational and methodological support for multidisciplinary on-farm research. Alternatively, existing institutions can combine their activities to form site-related research teams for which the staff of several institutions provides the expertise required. Such a model places heavy demands on site coordinators and complicates the administrative structure. It has, however, the potential for strong disciplinary backup and important feedback from on-farm research to policy makers.

Recent programs in cropping systems research in the Philippines[4] have tended to follow the latter model, but are primarily part of special projects rather than a general approach to the generation of agricultural technology by line agencies.

Conclusions

There has been a rapid increase in the availability of improved--often short duration--crop varieties, early crop establishment techniques, pest management alternatives, farm machinery, and supplemental irrigation. To be useful to farmers, these new technological components need to be carefully combined to fit the prevailing production environment. This requires a holistic approach to agricultural research that is oriented toward the combination of crop enterprises encountered on, or suitable for, the different land types in rice growing regions.

In formulating such an approach, it is best to avoid research methods that require complex computational and information processing techniques that must be applied by highly qualified, centrally located researchers. Cooperation with representatives from national research organizations in South and Southeast Asia (Cropping Systems Working Group, 1975) led to the formulation of a site-related cropping systems research methodology that focuses on the description and classification of the environment, on the design of improved cropping systems and their on-farm testing, and on methods for the formulation of production programs. Small multidisciplinary teams are now applying this methodology in more than 40 research sites in South and Southeast Asia.

---

[4] Such as in the land settlement projects in Agusan, Bukidnon, and Capiz and in the PCARR coordinated Bicol Agricultural Research Complex programs.

A remaining challenge is that of adjusting the
institutional structure to the requirements for site
related on-farm research. It needs to be addressed with
renewed vigor if agricultural researchers are to fulfill
their obligation to the farmer.

References

Baker, E. F. I., and D. W. Norman. 1975. Cropping sys-
    tems in Northern Nigeria. In International Rice
    Research Institute, Proceedings of the Cropping Sys-
    tems Workshop, March 18-20, 1975. Los Banos,
    Philippines. pp. 334-361.
Cady, F. B. 1974. Experimental strategy for transfer-
    ring crop production information. Cornell Univer-
    sity Paper No. BU-502-M, Ithaca, New York.
Carangal, V. R. 1977. Asian cropping systems network.
    In Symposium on Cropping Systems Research and Devel-
    opment for the Asian Rice Farmer. IRRI, Los Banos,
    Philippines.
Cropping Systems Working Group. 1975. Second cropping
    systems working group meeting, November 3-8, 1975,
    Indonesia. IRRI, Los Banos, Philippines.
Cropping Systems Working Group. 1976. Third cropping
    systems working group meeting, February 16-18, 1976,
    Thailand. IRRI, Los Banos, Philippines.
Cropping Systems Working Group. 1977. Fourth cropping
    systems working group meeting, September 20, 24-25,
    1976. IRRI, Los Banos, Philippines.
Food and Agriculture Organization of the United Nations
    (FAO). 1971. Soil survey project, Bangladesh
    agricultural development possibilities. AGL:SF/PAK
    6, Technical Report 2. FAO, Rome.
Gomez, A. 1977. Cropping systems approach to production
    programs. The Phillipine experience. In Symposium
    on cropping systems research and development for the
    Asian rice farmer. IRRI, Los Banos, Phillipines.
Harwood, R. R. 1976. The application of science and
    technology to long range solutions: multiple crop-
    ping potentials. In S. S. Nevin and M. Behar, eds.
    Nutrition and agricultural development. Plenum, New
    York.
Houser, G. F. 1970. A standard guide to soil fertility
    investigations on farmers' fields. Soils Bulletin
    No. 11. FAO, Rome.
International Rice Research Institute. 1974. An agro-
    climatic classification for evaluating cropping
    systems potentials in Southeast Asian rice growing
    regions. Los Banos, Philippines. 10p.
International Rice Research Institute. 1976. Annual
    report for 1975. Los Banos, Philippines. 470p.
Laird, Reggie J. 1968. Field technique for fertilizer
    experiments. Research Bulletin No. 9. El Batan,
    CIMMYT, Mexico.

Magbanua, R, D., N. M. Roxas, and H. G. Zandstra. 1977. Comparison of the turnaround period of the different groups of patterns in Iloilo. IRRI, Los Banos, Philippines.

Price, E. C. and R. Barker. 1977. A preliminary evaluation of the time distribution of crop labor as a criteria for design and testing of new rice-based cropping patterns. Paper presented at the Symposium on Household Economics, May 27-28, 1977.

World Bank. 1975. Rural development: sector policy paper. World Bank, Washington, D. C.

Zandstra, H. G., K. G. Swanberg, and C. A. Zulberti. 1975. Removing constraints to small farm production: the Caqueza project. IDRC 0582, International Development Research Centre, Ottawa, Canada.

Zandstra, H. G. 1976. Workshop on environmental factors in cropping systems. April 9-10, 1976. IRRI, Los Banos, Philippines.

Zandstra, H. G. and V. R. Carangal. 1977. Crop intensification for the Asian rice farmer. Agric. Mech. in Asia. Summer 1977 issue p. 21-30.

Zandstra, H. G. and E. C. Price. 1977. Research topics critical for the intensification of rice-based cropping systems. Paper prepared as background paper for research programming meeting on Cropping Systems Program at IRRI. Los Banos, Philippines.

Zandstra, H. G., K. G. Swanberg, C. A. Zulberti, and B. L. Nestel. 1979. Experiencias en desarrollo rural, proyecto Caqueza. International Development Research Center, Bogota, Colombia. 386p.

7
Motivating Small Farmers to Accept Change*

Peter E. Hildebrand

     This title suggests that small farmers do not accept change at rates which are considered adequate. Adequate could be defined in any of several ways, but it is not necessary to define it for our purposes. That these farmers are not changing their technology as rapidly as larger, commercial farmers is evident and will not be discussed either. Rather, presented here is an interpretation of the reason small farmers in developing countries do not accept changes in their current technology at rates which scientists, extensionists, politicians, academicians, bureaucrats, or others deem adequate. In addition, changes are proposed which can significantly modify this rate of acceptance. Admittedly, some of the suggested changes may well meet with the same resistance small farmers exhibit when presented with new ideas that would drastically modify their way of thinking and working.

     First, it is necessary to define some terms which must be used but which are vague or carry several connotations. The term "small farmer" will mean all farmers, regardless of the size of their holdings, who are not primarily commercial farmers, and most of whom in developing countries still use predominately traditional technology. Since we are concerned in this conference with technology, this is a much more utilitarian definition than one limited to size. Appropriate, as used in "appropriate technology," is necessary and desirable to use, but it is not used in the accepted or most commonly understood context. Appropriate technology will mean that technology (or change) which: 1) can be put into practice immediately and under farmers' present agro-socioeconomic conditions and 2) is acceptable to target farmers. The first criterion is a necessary though not

---

*Reprinted from *Agricultural Administration*, Vol. 8, 1981, by permission of Applied Science Publishers Ltd.

sufficient condition to be "appropriate"; the second reflects the difference between a third person's interpretation of farmers' agro-socioeconomic conditions and the farmers' own interpretation of the same things. In other words, it reflects the farmers' thinking and not macro or imposed micro considerations as interpreted by outsiders. "Agro-socioeconomic conditions" are all those agro-climatic, economic, social, cultural, or infrastructural factors or constraints which condition whether a farmer needs, desires, or can adopt any given change.

This discussion commences from the premise originally proposed by Schultz, and is widely, though not universally, accepted: small farmers are efficient in the utilization and allocation of available resources among known technologies if they have been farming under stable conditions for some time. As we are, by design and purpose in this conference, concerned with farmers who are not changing their production methods, this premise should include most of those farmers. This implies that small farmers will and do accept change when the available resource base changes or new and appropriate technology becomes known. Otherwise, they could not be efficiently adjusted to alternatives they now have. But it is important to understand that this efficient adjustment is in terms of the farmers' own understanding and interpretation of their situations, and it is not necessarily efficient according to the perceptions of well meaning but incompletely informed third persons. Since it is not third persons in a free society who make choice of technology and resource allocation decisions, it is evident that farmers' actions need not reflect third person solutions unless they are based on a nearly perfect conception of the farmers' situations.

A second characteristic of small farmers gradually being recognized is the high degree of location specificity of their agro-socioeconomic conditions. In commercial agriculture, the tractor and a strong capital base are effective homogenizers of what is otherwise a complex milieu. To persons who are trained or accustomed to being able to produce widely acceptable tractor-based technologies, this characteristic represents a strong barrier which hinders their effectiveness in producing usable and acceptable results for small farmers. But it is also a characteristic that must be considered explicitly in any technology developing system if it is to produce technologies which small farmers will be motivated to accept.

If small farmers are not changing their production methods because they are not being offered appropriate technology when so many people are working to produce it for them, what is the problem? If it is agreed that small farmers are efficient in the allocation of their resources to known and appropriate traditional technologies, it means they have been motivated in the past to

accept change. Hence, the problem is not one of motivation, as such. Rather it is one of offering "changes"
which are not appropriate as perceived by the farmers
themselves. It makes no difference to a farmer how a
third person views any specific technology. If he himself does not feel it to be appropriate, he is not going
to be motivated to accept it.

In turn, the problem stems from several different
areas. First, most top level technology "generators,"
who are agriculturally trained and "product" oriented,
work on experiment stations or in other highly controlled
conditions where they consider only a limited number of
variables. Second, most of the "transfer mechanism" generators, who are trained in the social sciences and are
"cause" but not product oriented, struggle with the vast
quantity of variables which condition acceptance or rejection of technology at the farm level. Finally, there
are the "goal" oriented agricultural economists in the
middle complaining that the agricultural scientists do
not consider enough of the variables of their work, but
ignoring the pleas of the social scientists who claim
that including just the quantifiable variables is not
sufficient either.[2] It is little wonder that the poor
extension or "change" agent has little to offer small
farmers even though he may be supported by an elaborate
experiment station and extension network manned by high
level technicians. It is even less amazing that small
farmers are not motivated to accept many changes that
come out of such a system.

ICTA Technology Development System

New technology development systems oriented toward
small farmers are being written about and discussed, and
a few are in operation. One which has shown promise and
is in use within a functioning national institution is
that at ICTA (Institute of Agricultural Sciences and
Technology) in Guatemala. This system has been developing over the last five years and is still changing as
needed modifications are visualized. It is not perfect,
but it has been found to have some valuable characteristics and is being used as a model in some other countries. Its most critical characteristics are briefly
sketched below.

---

[2] This picture is complicated further because agronomists
work primarily with soils and plants which they are convinced are the most important components of agricultural
production; sociologists and anthropologists work with
farmers who for them are obviously the most important
component; and economists work with desks and computers
studying means of achieving specified (and frequently un-
realistic) goals.

A work zone is defined, insofar as possible, on the basis of an area in which the majority of small farmers follow a similar traditional agricultural system; or in other cases, it may be the confines of a land reform projest where most of the (artificially created) farms are quite similar. A team composed of social scientists and the agricultural technicians assigned to the zone surveys the area to determine what the farmers do, how they do it, and why they do it that way (that is, define the agrosocioeconomic conditions of the area). This team jointly analyzes the results of the survey and makes recommendations concerning the technology to be developed. Technology validation and generation is carried out both on experiment stations (about 20 percent of the work) and on the small farmers' own farms (about 80 percent). This work is divided into three general levels. The commodity programs (those identified with a commodity such as maize, beans, swine, etc.) conduct highly controlled trials on the stations and a few farms in the area. A technology testing team (the technicians assigned to the zone) conducts technical trials under the supervision of the commodity programs on a much larger number of farms and acts as a means of extending the exposure of the materials and practices throughout the zone. The most promising technologies are then submitted to agroeconomic trials to help the team evaluate them further.

Ideally, the trials and evaluations through this stage are based on the technicians' understanding of the farmers' needs and criteria as obtained from the survey and from farm records which are initiated immediately following the survey. But, even though the technicians live in the area and work on the farmers' own land, they cannot make the final decisions as to the "appropriateness" of the technology even after passing it through this exhaustive system. Therefore, the most promising technologies are passed on to farmers for their own evaluation. Here the farmers pay for inputs and furnish labor, and the product is theirs. ICTA technicians obtain what information they can from these farmers' tests, but the farmers do the evaluation. The year following these tests by the farmers, ICTA makes a follow-up survey of the same farmers to determine whether they have adopted the technology, to what degree, and if not, why. If a sufficient number of the collaborators from the year before have adopted it of their own accord over a signifcant part of their own land, it is considered "acceptable" and is then turned over to the extension service as "appropriate technology" for those farmers who use that same traditional agricultural system.[3]

---

[3] In Guatemala, the extension service is separate from the technology generating institute. Ideally, these two functions should form a continuum within a single entity.

One of the strengths of this technology generating system is the use of multidisciplinary teams to make the agro-socioeconomic studies of each new zone of work and to aid in the evaluation and interpretation of results. For the survey, usually five social scientists (among them can be anthropologists, sociologists, economists, or agricultural economists) are paired with agricultural scientists (among whom may be found both plant and animal technicians in entomology, breeding, pathology, physiology, etc.). Besides changing interviewing partners every day to reduce interviewer bias and increase cross-disciplinary interchange, the group meets each night to discuss the day's findings, make preliminary interpretations, and modify the questionnaire if necessary. In order to be able to understand and interpret the small farmers' agro-socioeconomic conditions, it is necessary to consider all the factors which have an influence on what they do and can do. Hence it requires a multidisciplinary team each contributing his or her own specialty but all subordinating to the common objective: to understand what the farmers are doing, why they are doing it that way (how they have adjusted historically to their agro-socioeconomic conditions), and what is required in any new technology (proposed change) if it is to be accepted on a large scale.

The integrated multidisciplinary concept continues beyond the survey. The agricultural technicians on the team help the technician from socio-economics who is assigned to the team in the collection of farm record data and who, in turn, helps in the field trial work. Because this team lives and works in the zone and because the work is almost exclusively on farms, the technicians have a great deal of contact with the farmers in the area and continue to learn about their conditions both because of dialogue with them and because they are planting under farm conditions. Hence, they are able to obtain a very good understanding of the agro-socioeconomic conditions of the farmers in the area.

The System's Weakness

But there is still a weakness in the system. In the original organization of ICTA, the commodity programs were given the primary responsibility for increasing the production of their commodities. Though this concept predated the use of the multidisciplinary teams, it has persisted. As a result, even though multidisciplinary teams with a good understanding of the local conditions exist in each of the zones, they do not yet exert sufficient influence on the projects they carry out. Rather, they function in support of the commodity programs. Consequently, project orientation is not primarily in the hands of the personnel who best know each zone but in the hands of the commodity programs that have national responsibility and cannot be expected to have an intimate knowledge of each location.

The National Agricultural Research Program (PNIA) in Honduras, which is patterning its reorganization partly after the ICTA model, has seen the weakness just described and is organizing so that the multidisciplinary teams in each region have the primary responsibility for orienting technology development. This modification should also be made at ICTA. This type of reorganization need not affect the strength of the commodity programs which must have top level scientists to be able to respond to the need of widely different conditions throughout the country. But it will have to affect the concept of who supports whom within the Institute. Instead of conceiving that the technology testing teams, soil management, and socioeconomics support the commodity programs, it should be that soil management, socioeconomics, and the commodity programs support the resident multidisciplinary teams in each zone.

Organizing along these lines will obviously infringe on the concept of specialization which is traditional in agricultural research organizations. The principal requirement will be the need to upgrade the training of the people who make up the multidisciplinary teams. At present in ICTA, the technology testing teams in each zone include only university graduate or lower level personnel and none with graduate degrees (except for the Regional Directors who are in charge of several zones and whose function is largely planning and administration). Honduras, on the other hand, is placing some of its top researchers at the regional team level. If the commodity programs where the top people are now placed in ICTA are to respect the orientation coming from the zonal teams, it will be necessary not only to upgrade the level of training of these teams, but also to change the connotation which multidisciplinary work carries in many parts of the world, i. e., work done by undertrained generalists who have no strength in any discipline. As opposed to this non-disciplinary concept, a multidisciplinary team should be composed of people who are strong in their own field and who have enough confidence in their own work and enough respect for other fields that they do not feel the need to defend themselves from others and are not afraid to make contributions in fields other than their own.[4]

Persons with this type of training and inclination are very scarce and will need to be produced in large numbers. The first intent along this line of which the author is aware was the Cornell/CIMMYT program, supported

---

[4] See the appendix for some additional comments on multidisciplinary team efforts.

by The Rockefeller Foundation, that produced most of the group now working in PNIA in Honduras. Other programs of similar nature will have to be initiated, but in the meantime, great advances can be made even with the type of personnel now being used at ICTA in the multidisciplinary teams.

Summary

In summary, it should be repeated that the resistance of small farmers to accepting change is not one of motivation but rather one of not having technology available which is appropriate from these farmers' own points of view. Because of the location specificity of the agro-socioeconomic conditions of small farmers and because they are not subject to the homogenizing influence of tractors and capital, it is a much greater challenge to develop technology which they will be motivated to accept than it is to develop technology for commercial farmers. The most efficient way is by means of strong multidisciplinary teams who live and work in each area and who orient the technology development work undertaken for the small farmers in their zone. This implies a drastic change in the traditional role of many scientists now working on technology development and probably will meet with no small amount of resistance on their part. It may well be that in another, future conference on small farm technology, one of the papers will be titled, "Motivating Scientists and Technicians to Accept Change."

Appendix

Comments About Multidisciplinary Team Efforts

Individual and some collective action is being taken to bridge the differences generated by traditional scientific training in order to facilitate multidisciplinary efforts. Examples with which the author has had recent contact follow. Christine Gladwin is an agricultural economist who uses a methodology much more akin to anthropology than economics; Richard Harwood, an agronomist, found it necessary to combine his field with economics and sociology in order to bring acceptable rice technology to parts of Asia; Robert Werge is an anthropologist who is working in the field of agronomy to help the International Potato Center develop technology for this crop; and Daniel Galt, an agricultural economist, is actively engaged in crop trials in Honduras. Examples of their work are listed in the references.

All of the above researchers have two things in

common that are critical to the development of an efficient and functioning multidisciplinary team. They are
well trained in their own fields, but they also have a
working understanding of and are not afraid to make contributions in one or more other fields. This is a necessary characteristic of persons working on multidisciplinary teams. But alone, it is not sufficient. It is also
required that the team members not feel the need to
defend themselves and their field from intrusion by
others.

Another feature of a successful multidisciplinary
team is that all members view the final product as a
joint effort in which all participate and for which all
are equally responsible. That means each of them must be
satisfied with the product, given the goals of the team,
and be willing and able to defend it.

Returning to the generation of improved technology
for small traditional farmers, the team members must all
be product oriented, not just the agronomists.[5] Also,
all the team members must be willing to consider a wide
range of variables and constraints and not leave these
worries only to the anthropologists or sociologists.
Third, all members must be willing to spend some desk
time considering alternatives and their consequences on
the clients' goals and not leave this part of the task
just to the economists. The agronomists should be capable and willing to criticize the economic or social
aspects of the work, and the social scientists should be
willing and able to criticize the agronomic aspects. In
turn, these criticisms should be used to improve the
product so that all can be satisfied with the final
result.

Failures of multidisciplinary efforts frequently
have resulted because the teams were organized more as
committees that met occasionally to coordinate efforts
but in which the crop work was left to the agronomists,
the survey to the anthropologists, and the desk work to
the economists. In these cases, there is not a single
identified product but rather several products or reports
purported to be concerned about the same problem. Perhaps the most critical characteristic required to achieve
success of a multidisciplinary team is identification
with a single product in which all participate. The
product can be complex and involve a number of facets,
but it should result from the joint effort of the whole
team and not contain strictly identifiable parts attrib-
utable to individual team members.

In ICTA, the agronomists (who outnumber the social
scientists by about thirty to one) are concerned about

---

[5] Product, as used here, refers primarily to the technology produced and not the commodity itself.

there being too much influence by the socio-economic
group in the work at the farm level. This is manifest in
a certain resistance by the agronomists to identify too
closely with the farmers (even with those on whose land
they conduct trials). It also surfaces with respect to
evaluation of technology. The agronomist is much more
comfortable if a final evaluation follows the farm trial
phase of the work where it is the technician who makes
the evaluation. The technician then decides if a technology is "good." If the farmer evaluates this "good" technology and does not accept it, then the technician considers it a problem for the extension service, of poor
infrastructure, of low prices, or of lack of initiative
on the part of the farmer himself, but it is not a problem for the agronomist who has produced what he considers
to be a "good" product. In this situation, evaluation by
the farmer is equated with influence by socio-economists
who would tend to take into consideration more variables
including the present weaknesses in infrastructure, the
price level, the farmers' capabilities, etc., in the
development of a technology so that the product of the
team's efforts could be used immediately without the need
to await development of other facets of the sector. In
other words, in ICTA we have not yet completely identified
the kind of product we are to produce.

Even though we are a long way down the road, more
needs to be done at ICTA to make the multidisciplinary
teams and the efforts of the entire Institute more
efficient. The top management of the Institute (all of
whom are biological scientists) agree that socio-economics
must contribute directly to the generation of agricultural
technology, a concept with which we fully concur. On the
other hand, because of their own traditional training,
they also tend to be apprehensive about too much influence
from socio-economics and therefore are sometimes hesitant
to provide the kind of support which could enhance the
efficiency of the multidisciplinary teams much more
rapidly. Hence, another critical characteristic of a
successful multidisciplinary team effort is the conviction
of management and its understanding, dedication, and
support of the concept. Support at this level is required
in order to counteract the traditional resistance initially found at the field level.

A final necessary component for creating successful
multidisciplinary teams is a long run stability of the
government and/or its policies, so that management and
staff of national institutes who are expected to develop
technology for small traditional farmers, and for which
multidisciplinary teams are required, have time to work
out the details so they can function effectively.

References

Contreras, Mario Ruben, Daniel Lee Galt, Samuel Cephas
    Muchena, Khalid Mohamad Nor, Frank Byers Peairs, and
    Mario Santos Rodriguez P. 1977. An interdisciplin-
    ary approach to international agricultural training:
    the Cornell-CIMMYT graduate student team report.
    Cornell International Agricultural Mimeograph 59,
    Ithaca, New York.
Fumagalli, Astolfo and Robert K. Waugh. 1977. Agricul-
    tural research in Guatemala. Presented at a Bellagio
    Conference in October 1977, ICTA, Guatemala.
Galt, Daniel Lee. 1977. Economic weights for breeding
    selection indices: empirical determination of the
    importance of various pests affecting tropical maize.
    Ph.D. dissertation. Cornell University, Ithaca, New
    York.
Gladwin, Christina. 1976. A view of the Plan Puebla:
    an application of hierarchical decision models.
    American Journal of Agricultural Economics, Vol.
    LVIII, No. 5: 881-887.
Harwood, R. R. 1975. Farmer-oriented research aimed at
    crop intensification. In International Rice Re-
    search Institute, Proceedings of the Cropping Systems
    Workshop, March 18-20, 1974, Los Banos, Philippines.
    pp. 12-32.
Hildebrand, Peter E. 1977. Generating small farm tech-
    nology: an integrated, multidisciplinary system.
    An invited paper prepared for presentation at the
    12th West Indian Agricultural Economics Conference,
    Caribbean Agro-Economics Society, April 24-30, 1977.
    Antigua, W. T.
Schultz, Theodore W. 1964. Transforming traditional
    agriculture. Yale University Press, New Haven and
    London.
Secretaría de Recursos Naturales. 1978. Agricultural
    research in Honduras. Tegucigalpa, Honduras.
Stevens, Robert D. (ed.). 1977. Tradition and dynamics
    in small farm agriculture. Iowa State University
    Press, Ames, Iowa.
Waugh, Robert K. 1978. Research and the promotion of
    the use of technology. Symposium of the American
    Society of Agronomy. International Agronomy Division
    A-6 and the Extension Education Division A-4,
    December 3-8, 1978, Chicago.
Werge, Robert W. 1978. Social science training for
    regional agricultural development. Presented at the
    meetings of the Society for Applied Anthropology,
    Merida, Mexico.

8
Indonesian Cropping Systems Program

Jerry L. McIntosh

Objectives

In a developing country it is difficult for farmers to gradually adopt new technology as it is made available by research scientists. This is why production programs are so common in these countries even for the introduction of single component technology like new varieties, insecticides, and fertilizer recommendations. The introduction of new cropping patterns may take much longer and be infinitely more complex. This is especially true in irrigated areas where farmers cannot easily modify their cropping patterns without conflicting with their neighbors. For example, in fully irrigated areas we are sure from our cropping systems research that farmers could grow two crops of IR 36 rice and a soybean crop in one year. To do this, the first rice crop must be transplanted as soon as the water arrives or direct seeded before the arrival of the irrigation water. However, if one farmer plants early or uses an early maturing variety of rice while his neighbors follow their traditional practices, his rice will almost certainly be destroyed by rats or birds. Later, if he tries to plant soybeans after two crops of an early maturing variety of rice, his crop would likely be destroyed by flooding. His neighbors would still be growing their second crop of lowland rice. In this situation, even research is difficult to conduct. Consequently, insufficient research and difficulties in implementation impede cropping intensification.

Other examples of under use of lands are numerous. In Indonesia, the vast areas of tidal swamps and upland rainfed lands in Sumatra and Kalimantan have considerable potential for crop production. Presently, however, they are mostly covered by forests of *Imperata cylindrica*. In some places, new settlements have been started through the transmigration programs. Considerable research is needed to develop appropriate cropping patterns that are agronomically and economically sound for these areas.

The research must be integrated to include all components of the production system and at the same time provide for extension and marketing problems that arise with implementation.

The land use in Indonesia may be intensified and the area of production extended. The easy research problems for crop commodities and related fields have received considerable attention. Now our research must be directed to solving the problems that farmers face in their fields and integrated to include the scope of secondary problems that arise.

The overall objectives of the cropping systems research program may be summarized as follows:

The first is to increase food production by increasing total cropped area and productivity per hectare. This includes developing viable cropping systems for new lands, using more intensively present cropland, including interplanting food crops in estate crops such as rubber, oil palm, coconut, sugar, etc., and amending and maintaining soil fertility.

The second is to increase employment opportunity by increasing the opportunity for labor. This is accomplished by spreading out the time for planting and harvest, expanding the total area in production, and concomitantly increasing agribusiness.

The third objective is to improve the small farmers' bargaining position by increasing the frequency of harvests and minimizing the need to borrow (which may include items other than money).

The final objective is to facilitate institutional interaction and implementation of research findings.

Selection of Target Area

The objectives of cropping systems research cannot be met if the research is not implemented. The research must fit within the framework of the government and meet policy and developmental needs. If this is not the case, implementation will be difficult. Consequently, target areas for research must be carefully selected. Criteria have been developed as guidelines for selecting target areas for cropping systems research. The order of priority will depend upon the extent of government participation in food production activities. The criteria are:

1) Critical areas in terms of food shortages and governmental designation.
2) Large areas having similar soils and climate.
3) Feasibility of intensifying cropping patterns based on prior evidence.
4) Availability of markets and infrastructure.

These criteria are simple and straightforward. There are many sources of information that may be useful to administrators and scientists in making decisions to

concentrate a research program within a selected target
area. The availability of information varies from region
to region within Indonesia and from country to country.
The outline contained in Appendix 1 has been helpful in
gathering and making use of available information in
Indonesia. This outline is not intended to replace institutional land use planning activities but to help
cropping systems agronomists make use of information that
is usually readily available.

Cropping Systems Research and Development in Selected Target Areas

The objectives of cropping systems research may
appear overly idealistic and unattainable. However, the
Indonesian cropping systems program has gradually evolved
a systematic plan of work for this kind of research in
selected target areas. The interaction within the South
and Southeast Asian Cropping Systems Network has been
invaluable in this achievement. The systematic program
outlined in Table 1 is based on experience rather than
speculation within the Indonesian context. Other countries may not need to carry out all of the phases indicated and some may need more. Figure 1 shows how the cropping systems program fits into the CRIA[1] system in Bogor.
The program consists of a coordinated working group of
scientists from the various disciplines involved in the
program. The core staff emanates from the multiple cropping section of the Agronomy Division.

Site Selection and Description

These activities are carried out as soon as possible
after the target area has been selected. Most of the
data can be collected from secondary sources. The survey
and data collection teams should be interdisciplinary
groups of scientists and extension workers.

When selecting a site, the cropping systems scientist
should keep in mind that he cannot tackle all the conditions and problems that exist in a target area. A
brief survey and collection of secondary data from the
local government will usually provide sufficient information to enable the research coordinator to decide which
of the edaphological conditions he wishes to study. Further analysis of the data will permit confirmation or
rejection of a certain location as a possible research
site. The research coordinator must first stress what he
hopes to accomplish in the research. Then a logical
sequence of steps can be taken to ensure that the right
districts, sub-districts, villages, and farmers are

---

[1] CRIA is the acronym for the Central Research Institute
for Agriculture (Indonesia).

Table 1. Cropping systems research and development for selected target areas. CRIA, Bogor, Indonesia. July 1979.

Components	Phase I	Phase II	Phase III*	Phase IV	Phase V
Activity -	Site selection and description	Biological feasibility and evaluation	Design and testing of cropping patterns	Pre-production testing	Implementation
	I. Physical A. Soil taxonomy B. Rainfall distribution C. Irrigation D. Other climatic data II. Economic A. Agro-economic profile	I. Sequential testing on small plots A. Varieties B. Fertilizer response C. Crop combinations D. Other component technology II. Economic-farm recording A. Income B. Labor C. Market price III. Problem focused surveys	I. Partition of target area A. Water availability B. Soil capability C. Market accessibility II. Pattern design A. Farmers' design - monitor only B. Farmers' design - optimum mgmt. C. Improved design - low input D. Improved design - optimum mgmt. III. Testing--1000 m² plots‡	I. Researcher managed plots on 3-4 hectares A. Increase visibility and demonstrate potential II. Village level A. Identify biological and institutional large-scale production	I. BIMAS† type program for cropping patterns not commodities
Methodology -	Data collection and survey	Secondary data and small plots	Agro-economic evaluation in farmers' fields	Field level evaluation	Production program
Responsibility -	Research and extension	Research	Research	All relevant agencies	All agencies
Time frame -	Initial	Years 1 - 2	Years 1 - 3	Years 3 - 5	

*In this and succeeding phases, all planning must be coordinated by the Provincial Planning Agency (BAPPEDA).

†Production program for lowland rice.

‡Standardized data collection, data handling, data processing and reporting.

114

Fig. 1. CRIA functional framework.

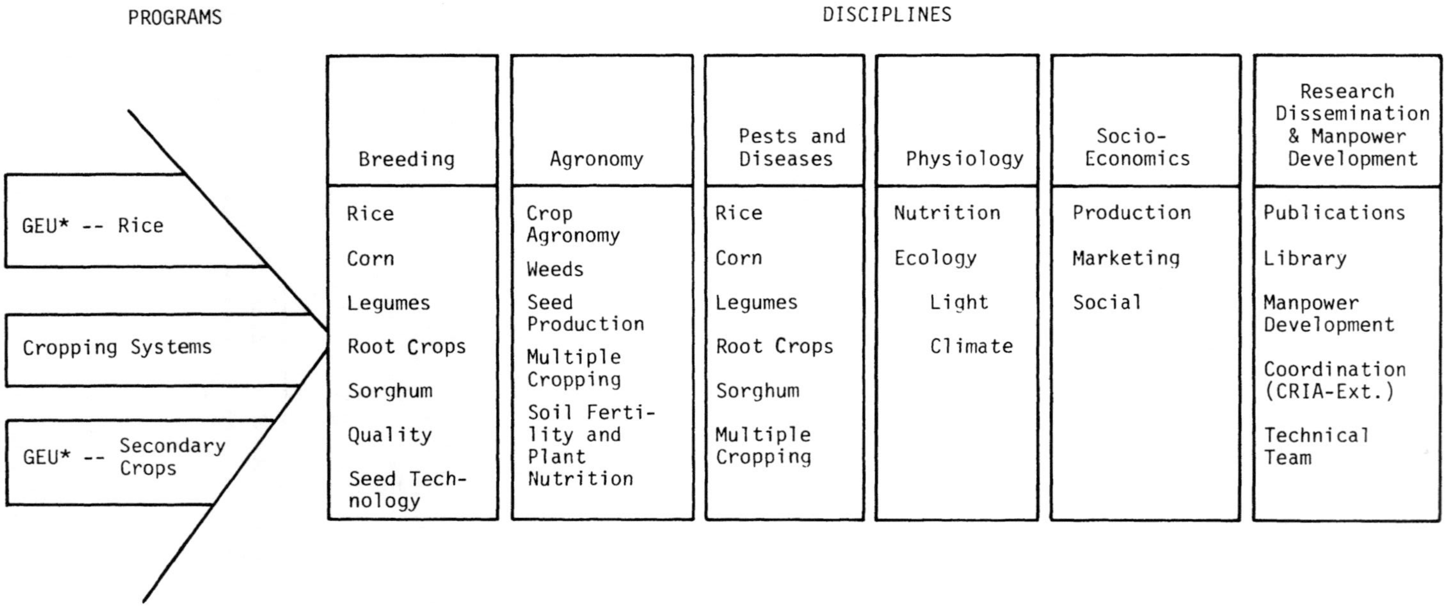

PROGRAMS

DISCIPLINES

Breeding	Agronomy	Pests and Diseases	Physiology	Socio-Economics	Research Dissemination & Manpower Development
Rice	Crop Agronomy	Rice	Nutrition	Production	Publications
Corn	Weeds	Corn	Ecology	Marketing	Library
Legumes	Seed Production	Legumes	Light	Social	Manpower Development
Root Crops	Multiple Cropping	Root Crops	Climate		Coordination (CRIA-Ext.)
Sorghum	Soil Fertility and Plant Nutrition	Sorghum			Technical Team
Quality		Multiple Cropping			
Seed Technology					

GEU* -- Rice

Cropping Systems

GEU* -- Secondary Crops

*Genetic Evaluation and Utilization

chosen. Appendix 2 gives an example of how this may be done.

Initially, secondary data can be collected to provide the physical and economic information needed for site selection. We may need more refined data for research purposes but most of all for transfer of technology to other places having similar agro-economic conditions. Below are two lists--one of physical factors and one of economic factors (determinants). These factors may be broken down in more detail as needed, but we have found there are many problems associated with collecting more data than necessary.

The physical factors are:

1) Soil taxonomy. This classification to the family level along with the usual analysis for soil fertility adequately describes the soil properties associated with plant growth, if the edaphological conditions explained earlier are taken into account.

2) Rainfall distribution. Monthly rainfall data collected over many years are available for most locations. We need to collect new data for the specific sites chosen. The long term data should be used not only for the average rainfall distribution but also analyzed for possible changes in the patterns and probabilities for starting and ending of the rainy season.

3) Irrigation. Length of time water is available and when it starts and ends.

4) Other climatic data. Solar radiation and temperature data should be collected if not readily available nearby.

5) Location and elevation.

The economic factor is: agro-economic profile. Details for this activity will be further described in Appendix 3. We prefer this term rather than baseline survey simply because it describes more accurately what is needed.

Biological Feasibility and Evaluation

These activities should be started as soon as possible after selection of the target area and research sites and continued as long as needed. Most of the agronomic studies can be conducted in small plots (3 x 5 sq. m) by the site coordinator and his assistants. Usually the team in each site consists of a team leader (agronomist), an assistant coordinator, and six field assistants. The assistant coordinator should be selected on the basis of need for a particular expertise in the site. If this is not possible, back-up expertise can be made available from the headquarters. The field assistants should be evenly divided according to biologic and economic research activities.

These small plot studies should be made at the time of the year and in the sequence (sequential testing) they would fit into the cropping patterns to be tested.

Many times adapted plant varieties are not available for new target areas. The cropping systems program should not become a breeding program, but some testing of new and introduced plant materials is appropriate.

In addition, fertilizer response curves for the macro nutrient elements are needed to determine the agronomic and economic thresholds. These should be uniformly carried out so that soil and climatic factors across the country (or region) may be better understood in relation to crop production.

Different intercrop combinations that are relevant must be evaluated just as for variety trials. Detailed studies concerning light, competition for nutrients, spacing, and economics may be more efficiently studied by scientists in the experiment stations.

Other component technology, such as guides for pest and disease management, must be developed.

Monitoring of the farmer cooperators and surrounding farm families must be started as early as possible. The data collection must be specific, the analyses quick, and the information used in design and testing of cropping patterns.

For research purposes we need to know the amount and distribution of the farmers' income and the extent to which government intervention is needed for implementation of research results. Also, the distribution of labor and the amount required for different patterns must be determined. Last, the selling and buying prices at the farmers' market level is needed on a weekly basis.

Rather than try to collect all the data in one large survey, it is better to focus on specific issues that may need study.

Design and Testing of Cropping Patterns

Cropping systems research can be complicated and confusing. Scientists must simplify the research approach as much as possible. This can be done by avoiding complex statistical designs that require sophisticated methods of data analysis. Examples of the methodology show how this can be done while taking into account ecological and socioeconomic factors that affect cropping patterns farmers use.

Even though a target area may fall within a single agro-climatic zone and edaphological class, there may be some variations which determine cropping patterns.

For lowland rice, the water availability or the length of time the soil can be flooded determines when and how many crops can be planted in one year. The classifications such as technical, semi-technical, and simple irrigation mean very little to cropping systems research. One target area in Indonesia is located in Indramayu, West Java. The area is characterized by relatively level topography, alluvial clay soils, three to four wet months with rainfall greater than 200 mm, and a long dry season. There are problems with water control--flooding during the rainy season and only partial irrigation during the dry season. The area was partitioned into four categories based on present conditions that are mostly dependent on water. These conditions would necessitate modifications or completely different cropping patterns. The bases for partition of the area into categories were:

Category I. Area with 10 months of irrigation water from October 1 to August 1 the following year.

Category II. Area with seven months of irrigation water from October 15 to May 15.

Category III. Area with five months of irrigation water from December 15 to May 15.

Category IV. Rainfed lowland (added later).

Soil capability was considered in selecting another target area that was an old transmigration scheme in Central Lampung. The area had been given a high priority for development by the government. The soil in the area was classified under the old system as red-yellow podzolic and similar to the soil of about 45 million hectares or approximately one-fourth of the land area of Indonesia. Furthermore, the rainfall which exceeds 200 mm for six months and falls below 100 mm for only three months is sufficient for year-round crop production, provided crops like cassava and cowpea are grown during the driest period. Unfortunately, the soil is low in inherent fertility and that contained in the organic component is soon lost after cultivation. Fertilizer inputs have not been available. As a result, this large agro-climatic zone is underdeveloped for agriculture. It is estimated there are about 20 million hectares suitable for agriculture but presently not used. Traditionally, farmers have used shifting cultivation and an extensive type of agriculture to circumvent the soil fertility problem. The transmigration schemes, however, are committed to a stationary agriculture. Farmers in older transmigration settlements have had difficulties in producing enough food to sustain their families. Our job is to develop cropping patterns and soil management practices that will enable the farmer to produce food for his family and have some surplus to sell. The original basis for partition of the area into categories was as follows:

Category I. Area with five months of irrigation.

Category II. Land opened from old *Imperata* fields.

Category III. Newly opened *Imperata* fields or secondary forests.

The research in Central Lampung in the upland areas is almost completed. Most of the research is now being conducted in new transmigration areas on newly opened land from either forested or *Imperata* covered lands. Much of the land is rolling to hilly and should not be

used for food crop production unless soil conservation
practices are used. Based on these conditions and our
past experience, we now propose to use the following
criteria for partitioning of the target area:
    Category I. Relatively level land on hilltops.
    Category II. Sloping land that must be terraced.
    Category III. Land from forests.
    Market accessibility must also be considered as a
dominating factor influencing cropping patterns suitable
for an area. In remote areas far from roads and markets,
food crops are grown mostly for subsistence. This is
especially true for crops like cassava which are difficult to store and transport. On the other hand, near
starch factories and good roads, cassava would likely be
the most valuable crop.
    For pattern design and testing, we will simply intro-
duce the reasoning that we have used to design cropping
patterns for testing in our selected target areas. Obviously, the priorities for different countries will depend upon the social and economic conditions that pre-
vail. Furthermore, we assume sufficient research in the
various disciplines (component technology) exists to
allow the cropping systems personnel to choose from among
a reasonably large selection of crops, techniques, and
management practices to meet the needs and objectives of
the research in the target areas.
    In selecting crops to be grown there are some crops
that are not suitable for inclusion in a cropping pattern to be tested in an area, even though the crop might
be suited agronomically. For example, in Indonesia sorghum grows well during the dry season when planted after
lowland rice. It is difficult to market at the present
time, however, and farmers will not eat it if they can
get rice or corn.
    Agronomic adaptation is obviously one important
consideration in selecting crops to be grown. The most
determining factor is rainfall and its distribution. In
Indonesia, food crops almost always receive the highest
priority. Of these, rice is the most highly valued crop,
and, consequently, it is planted if the rainy season is
long and sure enough. Corn would follow in terms of
value and length of the rainy season. Sweet potatoes
would be grown as a main food crop under conditions similar to corn in special areas where agriculture has
not developed. Cassava would be the most stable crop in
the drier regions or at certain times of the year.
Legumes, the kind depending upon the availability of
water, would be grown as catch crops. Some would be
retained for food and seed but most would be sold.
    Additional selection considerations are the market
and its potential. Most farmers grow crops primarily for
food for their families. Consequently, if they have
enough food (rice), they will not be likely to grow
another crop unless the marketing prospects are good.

This is true even for rice in Indonesia as a result of government policy to keep rice prices low. There is a concomitant effect on the prices of all food crops: crops which can be exported, such as cassava and corn, and those which can be processed, like soybean, mungbean, and peanut, offer a wider range of market potential.

To arrange cropping sequences, we took several facts into account. The average farm size in Indonesia is less than one hectare. In the outer islands, the holdings tend to be larger. Formerly, transmigrants received two hectares of land. They usually had enough labor to plant one-half hectare to food crops per year. The rest lay idle or grew up in *Imperata cylindrica*. Under these conditions there are certain things that the farmer intuitively considers. In a like manner, we must be able to interject ourselves into his situation in order to design effective and applicable cropping patterns. We have used the following guidelines in designing new cropping patterns for an area:

First, maximize stability in production. The concept is especially important in newly opened upland areas where the farmer must be self-sufficient. Under these circumstances, the farmer many times uses complex mixed cropping combinations with crop species ranging from early maturing legumes to cassava. For example, if there is some doubt about the amount of rainfall for rice, then perhaps early maturing corn should be interplanted with drought-tolerant cassava. After harvest of corn, the cassava may be interplanted with mungbean or cowpea to provide a more stable pattern.

Second, minimize labor. The area that a farmer cultivates depends mostly upon the amount of land he has or upon the amount of labor or power he has for land preparation. Usually a farmer with only hand labor can prepare about 0.5 hectare of land for planting at the beginning of the rainy season. Throughout the cropping season, weed control may become a constraint. Minimum tillage, relay planting, and continuous crop cover enable farmers to plant and manage a larger area for crops with the same amount of labor as for cropping patterns using monoculture and sequential plantings.

Third, distribute labor. The labor distribution inherent in multiple cropping systems is a useful attribute. Strip tillage and planting of intercrop combinations at intervals of two to four weeks enable a farmer to distribute his labor for land preparation for a given piece of land over a longer period of time. The harvesting time will also be spread out. Even under partially irrigated conditions where direct seeding of rice on moist aerobic soil is practiced, many times farmers interplant with corn. However, if this practice greatly increases the labor requirement, it may not be practical if the farmer has to hire labor.

Fourth, distribute capital inputs. Credit is

difficult to obtain by a farmer. Without government
assistance, the farmer has difficulty in buying seeds,
fertilizer, and insecticides. This is one of the primary
reasons farmers grow many kinds of crops in traditional
cropping combinations in upland agriculture in remote
areas. They plant what they have available. Again,
multiple cropping techniques similar to the farmers' may
be used to accrue the benefits of the farmers' systems.
But, the systems may have to be simplified to minimize
the randomness and diversity that prevent the farmer from
planting in rows, using specific fertilizers for higher
valued crops, and planting another crop soon after the
previous crop has been harvested.

Fifth, distribute harvest income. Frequent harvests
mean the farmer has money more often and, consequently,
is more likely to spend it for things he really needs.
It minimizes the need for borrowing money for inputs.
Again, the stability inherent in multiple cropping techniques is useful in this respect. There is a fine line,
however, between frequency of harvest and marketing efficiency. If the harvest is too small, the farmer may not
be able to afford to sell the product.

Research in the experiment stations contributes to
the pool of knowledge necessary to improve agricultural
production. Various components of cropping patterns can
be studied to understand principles of crop production
and interaction among plants. The latter may be described
as multiple cropping research to contrast it with traditional research in the various crop commodities. The
accumulative reservoir of information may be called component technology for cropping systems.

In developed countries where farmers may be well
educated and economically strong, the accumulated component technology may be sufficient to meet the needs of the
farmer. No further steps by researchers are needed. The
farmer is able to adapt the technology to meet his specific needs. In developing countries, however, where
farmers may be undereducated and financially weak, govern-
ments have initiated production programs to implement the
new technology. These are package programs which include
technology, credit, and availability of inputs. At first
these programs, such as Masagana 99 in the Philippines
and BIMAS in Indonesia, were for individual crop commodities. Recently, provisions have been made to include
cropping systems programs.

Before these programs for crop commodities and cropping systems reach the stage of implementation, they
should be preceded by research that approximates conditions at the farmers' levels of management. Production
programs are expensive and must be tailored to fit the
conditions that actually exist, if they are to be effective in increasing production. The first step entails
research in the farmers' fields under the management of
researchers to get some idea of crop performance and

production potential. If this looks promising, further
testing over a larger area is justified.

The final evaluation of cropping patterns should be
made through multi-locational trials conducted over the
target area under farmers' conditions and management, but
with and without removal of certain constraints such as
credit, seed, fertilizer, pesticides, and markets. Consequently, as an intermediate step between the farmer's
pattern and an imposed "improved pattern" we can study
the farmer's response to the removal of a set of constraints. Rather than imposing a cropping pattern upon
the farmer, we determine the kind he will use if the
agronomic inputs, credit, and markets are provided. This
assumes the farmer is not limited in technical know-how
(human technology). On the other hand, if the farmer
does not respond to the removal of the constraints but
continues to use his present cropping pattern and mis-
uses the agronomic inputs, we may conclude that he would
not be able to successfully participate in a production
program without a greater infusion of technical assistance by extension or, perhaps, simplified technology.

Three different cropping patterns were designed and
tested within each category for Indramayu and Lampung
beginning in 1975. Each trial was replicated three times
but by different farmers. The cropping patterns for each
category were not necessarily the same but were selected
on the basis of the same criterion. The criteria for
selection and the rationale for each criterion are as
follows:

Criterion A--Farmer's present cropping pattern.
Rationale--To establish a baseline check for comparison.

Criterion B--Farmer's choice of cropping pattern if
inputs and market constraints were removed. Rationale--
To evaluate the farmer's level of technical competence
and managerial skill and perhaps uncover hidden socio-
economic constraints.

Criterion C--Our introduced cropping pattern with
inputs and market constraints removed and technical
assistance provided. Rationale--To determine production
and economic potential and our ability to remove constraints.

A site coordinator, an agronomist, and an economist
were stationed in each target area. A field assistant
was put in charge of the work in each category and given
the additional responsibility of collecting all input-
output data. A system for collecting daily farm records
for all farm buying and selling activities was implemented in cooperation with 36 farmers in each target area to
get a larger base for socioeconomic evaluation.

The use of these criteria for design of cropping
patterns has been very helpful. It allowed us to be
objective and kept us from confusing cropping patterns
with cropping sequences. We do not get bogged down in
evaluating small differences in results from using

different species of legumes or varieties of rice in crop
sequences. These refinements are necessary but are the
kinds of research that are never finished. We have, however, been made aware of the severe economic stresses
faced by most Indonesian farmers. They simply do not
have much money they can use for inputs. If they do,
they are afraid to use it. This is particularly true for
farmers who have seldom worked with the Extension Service.
We feel we must develop low input patterns for new adopt-
ers. If the new technology is good and shows evidence of
being profitable, they will soon learn how to use more
inputs. We now use the following criteria for design of
cropping patterns.

Criterion A--Farmer's present cropping pattern
(monitor only). Rationale--To establish a baseline
check for comparison.

Criterion B--Farmer's cropping pattern with inputs
and optimum management. Rationale--To evaluate the
farmer's pattern without input and managerial constraints.

Criterion C--Our introduced pattern with low inputs.
Rationale--To induce the farmer to gradually try the new
technology.

Criterion D--Our introduced cropping pattern with
input and market constraints removed and technical assistance provided. Rationale--To determine production and
economic potential.

Preproduction Testing and Implementation

Cropping system research is problem oriented. Target areas are selected for in-depth research. For each
target area the activities include identification and
quantification of problems or possibilities, evaluation
of new technology in the field, preproduction testing
(pre-BIMAS testing), and transfer of technology to new
target areas.

At each step the Extension Service is involved.
Usually the research phase lasts for three years and the
involvement of the Extension Service and other provincial
services increase each year. In this way, the interface
between CRIA and Extension is increased and the involvement of the Provincial Planning Agency (BAPPEDA) facil-
itated. CRIA's targeted input ends with the implementation phase but, of course, the routine support continues.

APPENDIX 1

RATIONALE FOR INSTITUTIONAL RESEARCH PRIORITIES
AND CROPPING SYSTEMS RESEARCH

Agricultural scientists with less pragmatic inclination and more research orientation might disregard the
development needs and put more emphasis on personal or
scientific interests. Furthermore, the objective of the
research might be more devoted to in-depth study of small
differences or anomalies within an otherwise homogenous
target area. Fascination with details which do not pre-
clude uniformity of recommendations and cultural practices
should not become objectives in themselves. They should
not be forgotten but kept within perspective.

Indonesian agricultural scientists must provide the
technology and ideas for future agricultural development
activities. They must do research before they are re-
quested to provide answers. The stimulus for agricultural development should come from researchers rather than
the stimulus for research coming from development. In
this way, agricultural scientists will be able to serve
the country better, bring credit to themselves, and gain
support for their research organization.

Inventory of Resources

In addition to the traditional food crops research
activities and cropping systems research in target areas,
we need to develop a systematic way of arriving at priorities for adaptive agricultural research for all disciplines within CRIA. The subsequent research would pre-
cede development projects and even provide the initiative
for such projects. The first thing needed is an inventory
of natural resources and of the present agricultural
situation. The final stage in this approach is usually
the development of a "land use capability map." Such
maps have been developed for Indonesia. They are useful.
But for research, the logical sequence of information
that is needed for development of such maps may be more
valuable to the scientist than the final land use capability map. A series of maps presented in a sequence
from the edaphological classification of land, through
the physical determinants, and finally to the individual
food crops, would be more useful. It would help us see
where we are and what research might have more relevance
in all disciplines.

In edaphological classification of land, we attempt
to delineate distinct land areas that differ based on the
chemical and physical characteristics of the soil and
water environment, without reference to climate and other
overlapping factors such as slope or land form.

Some of the most important environmental factors
which determine the suitability of land for crop produc-

tion are soils, rainfall, elevation, and slope. The effects of environmental factors on land use capability vary depending upon the edaphological character of the land. These environmental factors may be looked upon as modifiers when used in combination with the edaphological map.

On a soils map, the soils delineated should be those whose characteristics necessitate different land management practices. For example, differences in inherent nutrient status would not be reason for differentiating between two soils unless one soil required unusual amounts of fertilizer for corrective treatment.

For the rainfall map, the classification described by Oldeman and the International Rice Research Institute (IRRI) Work Group are sufficient on a national scale. At the working level (district) bar graphs for rainfall distribution are more useful.

A biological classification in which altitudes between 500 M and 1,000 M are delineated would be sufficient for a national elevation map. These would correspond to the elevation above which cold tolerant rice varieties are needed (> 500 M) and the altitude above which wheat grows well (> 1,000 M). At altitudes higher than 1,500 M (another elevation may be more valid) the use of the land for food crops production is limited.

On a slope map, an average slope above which agricultural activity is limited is difficult to define. A slope of 15 percent has been considered the cut-off point for food crops production. Obviously, many times land with more than 15 percent slope has been used for crop production without any extreme problems with erosion. On Java and Bali where terracing is widely practiced for lowland rice, much steeper slopes are modified for use and the slope factor becomes almost irrelevant. This is an example of farmers modifying or removing physical constraints to crop production.

In development of land or research objectives within an area, the most significant data available are the present land use and information obtained from farmers. What exists cannot be disregarded. On a national scale, the following land use classifications may be useful: upland food crops; lowland rice (including rice grown in swamps and tidal areas); mixed *Imperata cylindrica* and brush land; forest (primary and secondary); and perennial estate crops.

The land use information delineated can be valuable in two ways. First, it is useful to relate land use (by distinctly different crops or vegetation which have different ecological needs) to a physical setting that can be characterized. Further breakdown by crops or species of plants provides the "standards" for evaluating land capability. They give some bases for modification of present land use or extrapolation of a particular kind of land use into new areas having similar agro-climatic conditions. Secondly, production figures for different food crop commodities from different areas of the country provide a basis of comparison. If production in areas with similar agro-climatic conditions differs greatly, we are provided with an ideal problem for applied and basic research projects that have relevance. We have rational bases for developing research priorities.

Interpretation and Decision Making

Use of Resource Maps

The combination of all the factors that affect crop production into one functional land use capability map (survey map) is difficult. It is not necessary to try. The Soils Research Institute has made these kinds of maps. They are available and are useful for many purposes. For an overview, the inventory maps described (scale of 1 : 2,500,000) are adequate. It may be useful to have more detailed maps of each major island group at a scale of 1 : 1,000,000.

Working maps, at a scale of 1 : 50,000 are needed for provinces or groups of provinces that may be treated as a unit. This would translate to 1 cm of map for each one-half kilometer of land and would provide sufficient detail for most agricultural purposes. Unfortunately, data in this detail are not available for much of Indonesia. However, enough data are available in detail to provide thorough agro-climatic descriptions of parts of many of the major agricultural areas. Furthermore, many surveys funded by the Directorate General of Transmigration and the Ministry of Public Works are detailed descriptions of forested and grass covered lands not yet investigated by agricultural researchers. These reports have been prepared by some of the best consulting firms available anywhere. The data in these reports along with the research and experience of CRIA staff are valuable resources. In combination with the survey maps, enough data are available to provide the interpretation and extrapolation needed for establishing national research priorities.

The usefulness of the large scale survey maps and working maps may be enhanced by considering just the relevant combinations. For example, a land use map of upland areas in combination with soil, rainfall, elevation, and slope maps, would be useful.

If we can identify certain upland crops (or cropping patterns) or perennial crops presently growing in one location, we might expect to find (or plan to grow) the crop in another location with similar agro-climatic conditions. The upland crop areas are the most complex.

For the swampy and tidal areas, more detail is needed than we have indicated in the survey maps for Indonesia. In many instances the delineation of factors

such as depth and nature of peat and acid sulphate are
not clear. Extrapolation of results from one area to
another is risky until we have more detailed information.
However, our work has been made easier by farmers who
have pioneered the development of some of these areas.
We should work with the pioneers first and then push into
the unsettled areas as we gain more information and experience.

Other Data Needed

The classification and inventory of physical data
are essential for the development of research priorities.
Unfortunately, many times the constraints to food production in Indonesia are more related to socioeconomic than
agronomic factors. Many times biological research scientists have been content to emphasize (or point out) this
problem but not go further and help find a solution. If
an economic constraint exists or is suspected, the scientist could make a significant contribution by documenting
the problem and suggesting ways to solve it. Many times
it is argued that crops like corn and sorghum are not
grown more often because farmers cannot make money growing them. If this is true, the sorghum agronomist would
make a significant contribution by helping the economist
document the costs of production and giving some idea of
a fair floor price.
Furthermore, the reservoir of germ plasm for different crops throughout the world is extensive and varied.
We need to characterize more precisely the kind of plant
materials needed for different cropping patterns in agro-
climatic regions throughout Indonesia. We can start by
collecting this information from scientists in the regions.
In this way we can begin to systematize the collection of
germ plasm from abroad for immediate evaluation and for
varietal improvement.

APPENDIX 2

SITE SELECTION IN TARGET AREA

R. H. Bernsten

Cropping systems research activities are designed to
accelerate agricultural development by increasing both
yields and cropping intensity. The program is field
oriented with almost all of the research conducted on
farmers' fields.
Four steps are involved in locating farmers' fields
in which the field trials are to be implemented. First,

a target area is identified which is a relatively homog-
eous agro-climatic area including several districts and
several thousand hectares. The cropping systems research
coordinator must decide which edaphological condition to
study, such as rainfed, irrigated (full, seven to nine
months, or five months), tidal, or swampy. Second, one or
several subdistricts are selected from among these districts that include a large area in the desired research
environment. Next, one or more villages characteristic
of each desired environment are selected. Finally, cooperating farmers are chosen in each village. The
decision criteria for proceeding from target area to
farmers' fields are discussed below.

Target Areas

    The selection of target areas for cropping systems
field research is based on four criteria. First, target
areas are usually regions identified by the government as
priority agricultural development zones. Second, the
area must be representative of a large agro-climatic zone
so that the research results will have widespread applic-
ability. Third, the environment must be of a type in
which the research staff believes there exists improved
agricultural technology so that with slight modifications
it will be possible to increase yields and cropping intensity. Finally, the target area must have some marketing
and infrastructural development or be in the process of
developing these facilities.

Subdistrict Selection

    In selecting the subdistricts, the primary consid-
eration is to identify an area which has a large number
of hectares of the desired land use type. The research
staff visits each district extension office and collects
secondary data for each subdistrict about the number of
hectares of rainfed, technical irrigation, semi-technical
irrigation, simple irrigation, annual crop upland, and
perennial crop upland. Based on these data, the subdistrict with the largest area of the desired land use
type is selected.

Village Selection

    The selection of the villages involves several con-
siderations. The research staff visits each of the
chosen subdistricts and collects from the extension
office the secondary data listed in Table 2.
    Once the secondary data are collected, a matrix is
prepared for each subdistrict with the village forming
the rows and the data forming the columns, as shown in
Table 3.
    After transforming the village secondary data to the

Table 2. Data required for systematic selection of
village sites.

Data	Purpose
Distance from main road (km)	To guarantee that the village is easily accessible.
Area in each land use class (ha)	To permit the selection of villages with a large hectarage in the desired land use class.
Relative area in each slope class (%)	To avoid villages with atypical topography.
Relative area in each soil texture (%)	To avoid villages with atypical soils.
Area planted to each crop, by month (%)	To identify current production level.
Population, by economic activity (number)	To determine importance of agricultural employment.
Rainfall by month for past 10 years (mm)	To determine number of months with 100 mm or more of rain and probability of less than 100 mm at beginning and end of cropping season.
BIMAS participants (number)	To determine the availability of credit and level of technology in the village.
Months during which irrigation water is available (% of area with less than 5, 6-7, 8-9, and 10 months or more of irrigation)	To identify areas with the respective irrigation regimes.
Draft animal population (number)	To determine the availability of draft power.
Tractor population (number)	To determine the availability of mechanical power.

Table 3. Cropping systems village selection data matrix.

District _____

Subdistrict _____

No.	Village	Distance (km)	Irrigation Tech & Semi-Tech.	Simple Tech.	Rain-fed	Upland Annual	Peren-nial	Slope (%) Flat	Rol-ling	18%	Moun-tain-ous	Soil (%) Clay	Silt	Sand	Cropping (%) LLR	ULR	C	CV	SB	PNT
		(1)	(2)	(3)	(4)	(5)	(6)	(7)	(8)	(9)	(10)	(11)	(12)	(13)	(14)	(15)	(16)	(17)	(18)	(19)
1.																				
2.																				
3.																				
.																				
.																				
15.																				
Mean																				

No.	Village	Yields (kg) LLR	ULR	C	CV	SB	PNT	Population Number Total	Male Adult	% Farmer	Gov't Program (%) Farmers* Bimas*	Immas*	Power Hectares per: Animal	Tractor
		(20)	(21)	(22)	(23)	(24)	(25)	(26)	(27)	(28)	(29)	(30)	(31)	(32)
1.														
2.														
3.														
.														
.														
15.														
Mean														

LLR = Lowland Rice   ULR = Upland Rice   C = Corn

CV = Cassava   SB = Soybean   PNT = Peanuts

*These are two government production programs, e.g., BIMAS is for lowland rice.

"data matrix," the mean value for each characteristic is calculated. These mean values taken together may be interpreted as a description of the "typical or representative village." To identify the village which is most representative of the population of villages, first the mean value for each characteristic is subtracted from the respective values associated with each village. This difference is the deviation from the mean for each characteristic. Next for each characteristic, the village with the smallest deviation from the mean is assigned the value of one, the village with the second smallest deviation is assigned the value two, etc., until all villages have been ranked in terms of deviation from the mean. Finally, after ordering all villages for all characteristics, each row (representing one village) is summed. This gives a single index value for each village. The village with the smallest index value will be most representative of the population of villages. Unless this village has some characteristic that precludes the establishment of a site there, it is selected as the research site.

A simple illustration of this procedure is shown in Tables 4, 5, and 6. In Table 4, a set of fabricated data is presented. Based on the mean values for each characteristic, the absolute deviations are shown in Table 5. Each village is then assigned a value of one to five for each characteristic to indicate its order of magnitude among the population of villages, as shown in Table 6. We see that village No. 4 has the lowest numeral value, so it is most representative of the five villages in terms of the 16 characteristics considered.

In this illustration, all characteristics are given equal weight, i.e., each contributes one-sixteenth to the "sum" index. Yet, if the researcher believes that certain characteristics should have a greater impact on village selection, it is possible to increase the relative contribution of such characteristics on the "sum index" by multiplying those items by any desired value. For example, by multiplying the rank-order value of characteristic one (distance), by five, it's weight in the final "sum index" would increase from one-sixteenth to five-twentieths.

Table 4. Characteristics of potential cropping systems village sites.

No.	Village	Distance (km)	Land Use (Ha)			Soil (%)			Cropping		(%)	Yield (t/ha)			Farmer popula-tion(%)	BIMAS members(%)	Power (ha/animal)
			Irrigated	Rainfed	Upland	Clay	Silt	Sand	LLR	C	CV	LLR	C	CV			
		(1)	(2)	(3)	(4)	(5)	(6)	(7)	(8)	(9)	(10)	(11)	(12)	(13)	(14)	(15)	(16)
1.	Maritengae	6	600	5,000	700	55	30	15	60	30	10	3.0	0.7	6.7	75	45	10
2.	Panca Rijang	10	4,000	1,000	600	50	20	30	70	20	15	2.8	0.5	5.4	63	33	15
3.	Branti	15	8,000	2,000	1,000	90	5	5	80	15	5	4.1	1.3	10.6	81	68	6
4.	Watang Pulu	7	3,000	100	2,000	75	13	12	68	25	7	3.4	0.8	8.4	68	60	21
5.	Dua Putue	4	600	900	6,000	85	5	10	75	5	20	3.5	1.0	9.0	74	50	9
	Mean	8.4	3,240	1,800	2,060	71	14.6	4.4	70.6	19	11.4	3.36	0.86	8.0	72.2	51.2	12.2

Table 5. Absolute deviation from the mean of each characteristic.

Village No.	Characteristic															
	(1)	(2)	(3)	(4)	(5)	(6)	(7)	(8)	(9)	(10)	(11)	(12)	(13)	(14)	(15)	(16)
1	2.4	2,640	3,200	1,360	16	15.4	0.6	0.6	11	1.4	0.36	0.16	1.3	2.8	6.2	2.2
2	1.6	760	800	1,460	21	5.4	15.6	0.6	1	3.6	0.56	0.36	2.6	9.2	18.2	2.8
3	6.6	4,760	200	1,060	19	9.5	9.4	9.4	4	6.4	0.74	0.44	2.6	8.8	16.8	6.2
4	1.4	240	1,700	60	4	1.5	2.4	2.6	6	4.4	0.04	0.06	0.4	4.2	8.8	8.8
5	4.4	2,640	900	3,940	14	9.5	4.4	4.4	14	8.6	0.14	0.14	1.0	1.8	1.2	3.2

Table 6. Rank-order of village characteristics for all villages in Kecamatan.

	No.	C h a r a c t e r i s t i c															
Village	1	2	3	4	5	6	7	8	9	10	11	12	13	14	15	16	Sum Index
1	3	3	5	3	3	4	1	5	4	1	3	3	3	2	2	1	46
2	2	2	2	4	5	2	5	1	1	2	4	4	4	5	5	2	50
3	5	4	1	2	4	3	4	4	2	4	5	5	4	4	4	4	59
4	1	1	4	1	1	1	2	2	3	3	1	1	1	3	3	5	33
5	4	3	3	5	2	3	3	3	5	5	2	2	2	1	1	3	47

APPENDIX 3

AGRO-ECONOMIC PROFILE OF THE SELECTED
CROPPING SYSTEMS SITE

R. H. Bernsten

Introduction

In order to design cropping patterns appropriate for new target area research sites, a preimplementation data collection effort is required. First, the data collected should comprehensively describe the selected village, including the physical, institutional, social, and economic environment. Second, the report should be not only descriptive but also designed to identify constraints to higher yields for specific crops, input intensification, crop intensification, and technologies which are characteristic of the alternative cropping systems strategies that are being considered for target area testing. Third, the agro-economic profile must be completed in a minimun of time, not exceeding two to three days per site. Fourth, the final report must be short, so it can be completed in a maximum of two weeks after returning from the field. Fifth, the data collection and report must follow a general framework that may be used at each new cropping systems site. This is necessary to reduce the time required for data collection and report preparation. In addition, the use of a general model will permit comparison of new sites to ongoing research areas. This will enable the researcher to evaluate the transferability of technologies found to be successful at old sites to the new sites.

The General Research Data Model

Data for developing the agro-economic profile should be collected from the source capable of giving the most accurate answer in a minimum of time. The required secondary data are usually available from such sources as the village office, Extension Service, Bureau of Central Statistics, Irrigation Office, the bank extending BIMAS credit, and input dealers. When the required data are not available from these sources, a key informant may be relied upon. Possible key informants include extension officers, village officials, village water officers, and a group of approximately 10 farmers assembled for the purpose of providing the information sought. This comprehensive set of data required for cropping systems design is listed in Table 7 by subject categories.

134

Table 7. Agro-economic profile data requirements by
          subject category.

Subject Category	Subject Category
Physical Environment	**Labor**
Rainfall*	Employment profile
Soil*	Population
Topography*	Off-farm employment
Land use by type*	Migration of agricultural labor
Experimental Base	**Farm Practices**
Variety trial	Wages
Fertilizer trial	Power
Pest surveillance	Input use
Demonstration plots	Yield constraints
	Varieties
Crop Situation	Planting decision rule
Hectares in each crop*	Input levels
Planting and harvesting dates*	Constraints to intensification
Yields*	
Current cropping pattern	**Prices**
Historical cropping pattern	Inputs
	Outputs (crops)
	Subsidies
Institutional	**Community**
Land ownership	Transportation
Tenure	Markets
Landless labor	
Support Services	
Credit	
Input sales	
Input availability and timeliness	
Irrigation system	

* These items should have already been collected before
choosing the village.

9
Farming Systems Research at ICRISAT*

B. A. Krantz

Farming systems research (FSR) involves a holistic approach to interdisciplinary systems research. Since this could include the synthesis of an unmanageably wide range of disciplinary activities, the FSR scientists first must survey and analyze the present setting, the natural and human resources, and the available research information in relation to future potentials and then must develop a sound approach in priority areas.

At ICRISAT we are concerned with the development of farming systems which would help to increase and stabilize agricultural production through the better use of the natural and human resources in the seasonally dry, semi-arid tropics (SAT). The objective of this paper is to discuss the setting and the present situation in the SAT as a framework for the conceptualization of the major problems involved, and the approaches and methodologies to be used in investigating alternative farming systems for the small farmer of the SAT. Some of the results obtained will also be presented for illustrative purposes.

The Setting

The SAT where precipitation exceeds the potential evapotranspiration for about 2 to 4.5 months per year (Troll, 1966) represents a diversity of soils, climates, and people. The area, which is home to about six hundred million people, is characterized by soils low in organic matter (0.5-0.8 percent) and fertility, and by undependable rainfall. Under these conditions, rainfed agriculture has failed to provide even the minimum food requirement for the rapidly increasing populations of

---

*ICRISAT is the acronym for the International Crops Research Institute for the Semi-Arid Tropics located in Hyderabad, India.

many developing countries in the SAT. Although the reasons for this are many, the primary constraint to agricultural development in the seasonally dry tropics is the lack of suitable technology for soil and water management and viable crop production systems.

In most regions of the SAT, the average annual rainfall would appear to be sufficient for one, or in many cases two, good crops per year. However, the rainfall patterns are erratic and undependable with frequent rainless periods even within the rainy season. The coefficients of variation of the monthly rainfall for June, July, August, September, and October are 57, 45, 52, 59, and 94 percent, respectively.

Alfisols and Vertisols are the two soil orders found in greatest abundance in the semi-arid tropical zone. Although Alfisols and Vertisols may occur in close association, their management requirements are distinctly different. The most striking example of this fact is the farmers' practice of cropping Alfisols only during the rainy season and cropping deep Vertisols only during the post-rainy season. The management requirements are related to differences in type and amount of clay, workability, moisture-holding capacity, and other associated characteristics.

The Alfisols (Ustalfs) discussed in this paper are fine, kaolinitic, isohyperthermic members of the family of Udic Rhodustalfs. The plant-available moisture storage in the root zone of these soils is usually less than 100 mm. The slopes of these soils range from 0.5 to 3 percent and erosion may be serious, particularly under conditions of inadequate crop cover. The soils are moderately weathered, with a base saturation of about 80 percent, which is dominated by calcium. The soils are low in organic matter, nitrogen, phosphorus, and often zinc. The potassium level is usually adequate and pH ranges from 5.8 to 6.7.

The Vertisols (Usterts) referred to in these investigations are fine calcareous, montmorillinitic isohyperthermic members of the family of Typic Chromusterts. The Vertisols are high in montmorillinitic clay (50 to 64 percent) and undergo pronounced shrinkage during drying, resulting in large cracks that close only during prolonged rewetting. These soils become hard when dry and sticky when wet. The slopes range from 0.5 to 3 percent and erosion is a serious problem, particularly under rainy season cultivated fallow. The soils are high in bases, including calcium, magnesium, and potassium, and the pH ranges from 7.5 to 8.6 percent. Under semi-arid tropical conditions, the soils are low in organic matter and are usually deficient in nitrogen, phosphorus, and sometimes zinc.

Because of the uncertainties and ever-present risk of droughts, farmers in the SAT have been reluctant to adopt the use of high yielding varieties, fertilizers, and other inputs characteristic of the Green Revolution in some areas. During the past 30 years, the population of many countries in the SAT has doubled; farmers have therefore attempted to double agricultural production. Since there has been no appreciable increase in per-hectare yields during this period, the result has been an increase in the areas devoted to crops. This increase is especially high in the SAT. Recent surveys in 84 districts of the SAT of India showed that 57.2 percent of the total areas of these districts were cultivated compared to only 44.6 percent for the country as a whole (Anon., 1970). In the Sholapur and Bijapur districts of India, which are composed mainly of Vertisols, the proportion of the geographical area presently cropped is 81 to 84 percent, respectively (Ryan, 1976). Thus, steeper and more erodible lands are being cropped and over-grazed and forest areas are being denuded causing permanent damage to vast areas.

People in the SAT depend primarily on agriculture for employment. Present production and income levels in most of these seasonally dry, rainfed areas do not fulfill the basic human needs. This situation is caused by low and unstable agricultural production due primarily to the lack of proper technology to manage the erratic and undependable rainfall. The people of the SAT have found through long and bitter experience that nature itself is so unpredictable that their system of farming is a hazardous way of life. In this setting and in line with the ICRISAT objective, the major goal of FSR is "to contribute to raising the economic status and quality of life for the people of the semi-arid tropics by developing farming systems which increase and stabilize agricultural production" (Krantz and Kampen, 1973).

Past approaches to alleviation of production problems in the SAT were:

1) Breeding of high yielding varieties.
2) Agronomic and fertilization studies on high yielding varieties.
3) Fallowing of deep Vertisols during the rainy season in an attempt to accumulate a moisture reserve in the soil profile.
4) Soil conservation by contour bunding.
5) Emergency programs to meet droughts and food crises.
6) Development of large irrigation projects.

Since water is the most limiting factor in crop production in the SAT, these approaches did not increase or stabilize crop yields appreciably (Kampen and Associates, 1974). This lack of increased per hectare yields in many developing countries has resulted in increased pressure on land, expansion of cultivated agriculture into marginal areas, overgrazing, deforestation, and severe soil erosion on vast areas of land. Thus, the land resource base is shrinking and the productive

capacity diminishing; this in turn increases the need for
more land. To break this vicious cycle, more stable
forms of land use which preserve and maintain the productive capactiy are urgently needed (Kampen and Associates,
1974).

As the FSR program at ICRISAT was being developed,
some major problem areas which appeared to need immediate
attention were:
1. About 18 million hectares of deep Vertisols in
India and millions of hectares in Africa were being clean
fallowed or being left to unproductive uses during the
rainy season. The low productivity of post-rainy season
crops grown on residual moisture seemed to indicate in-
efficient utilization of the water resources. The exposure of the fallowed soil to the impact of intense rains
has resulted in greatly increased soil erosion in spite
of present soil conservation measures.
2. In the Alfisol areas of the Indian SAT, tank and
well water was being used mainly on high water-requiring
crops such as rice and sugarcane. In the SAT where runoff and ground water is limited, very few research
efforts had been made to explore the question of how
limited water resources could be used to "back up" rather
than to replace rainfed agriculture.
3. In most of the Vertisol areas of the Indian SAT
and all areas of African SAT, there are few programs of
surface or ground water storage during the long dry
seasons even though water is so scarce that it often must
be carried long distances for domestic use.

The basic reasons for most of these problems appeared to be a lack of relevant soil, water, and crop management research. This research is essential for the development of viable soil and water management and utilization technology for the small farmers in the rainfed SAT.
Obviously, the solutions to these complex problems are
not simple and single component approaches cannot be
expected to work. Thus, it appeared clear that a holistic
approach to systems research on soil, water, and crop
management was essential.

Hypotheses and Concepts

Some of the hypotheses or concepts which formed the
basis for FSR approaches and strategies at ICRISAT were:
1. In the rainfed SAT, water is the most limiting
factor to production and all systems must be geared to
its optimum utilization.
2. Soil erosion is a serious problem in the SAT.
New soil and water conservation methods, which will also
increase yields substantially, are urgently needed.
3. In rainfed agriculture, where the only source of
water is rainfall, the watershed (catchment) is the
logical unit for investigating the optimum development
and management of the water and soil resource.

4. Runoff, erosion, infiltration, groundwater re-
charge, drainage, and other hydrologic factors do not
express themselves in small-sized experimental plots.
These factors can best be studied in watershed units.

5. The small subsistence farmers of the SAT are
dependent mainly upon animal power and human labor. No
rapid change in access to mechanical power is envisaged
nor does that seem desirable. Therefore, FSR should
optimize the use of these energy resources in trying to
develop viable technologies.

6. Improved equipment that is appropriate and low
cost is essential for implementing more efficient soil,
water, and crop management practices.

7. Many production and harvest problems encountered
by farmers will be realized by scientists only if research
is conducted on field-scale operational units.

8. Improved varieties, fertilization, and crop
management practices better utilize the available natural
and human resources and are essential ingredients to help
increase and stabilize production and improve the quality
of life for the people of the SAT.

The research strategy was: to simultaneously investigate single production components in depth and also
to integrate these components in a holistic manner in
systems research on an operational scale (Fig. 1); and to
investigate and test hypotheses and to develop approaches
and methodologies which would have wide application and
could be used by national programs to tailor the research
findings to their specific conditions (Binswanger et al.,
1976).

Requirements of Soil and Water Management Systems in the SAT

In planning improved soil and water management systems, the above mentioned characteristics of soil and
climate, as well as farm sizes, and the human, capital,
and power resources must be considered. Viewing these
characteristics, some of the specifications of an improved soil and water conservation and management system for
rainfed cropping areas would be as follows: avoid large
concentrations of water and large streams, except in a
protected grassed waterway; lead the water slowly off the
land in small streams uniformly spaced over the land
(watershed) so as to reduce erosion, increase water-intake
opportunity time, and provide drainage during prolonged
rainy periods, especially on deep Vertisols; provide year-
round protection against erosion, even during the occasional storms of the hot dry season; establish grasses
which are highly productive and palatable so as to provide nutritious forage for milk or draft animals and to
protect against erosion of the drainage way; in the drainage ways, use a combination of forage legumes and grasses
to minimize nitrogen requirements and provide more

140

Fig. 1. Organizational chart of the FSR program showing FS subprograms directly involved and the cooperation with the crop improvement, training, and economics programs at ICRISAT and cooperative national programs in the SAT.

THE FARMING SYSTEMS RESEARCH PROGRAM

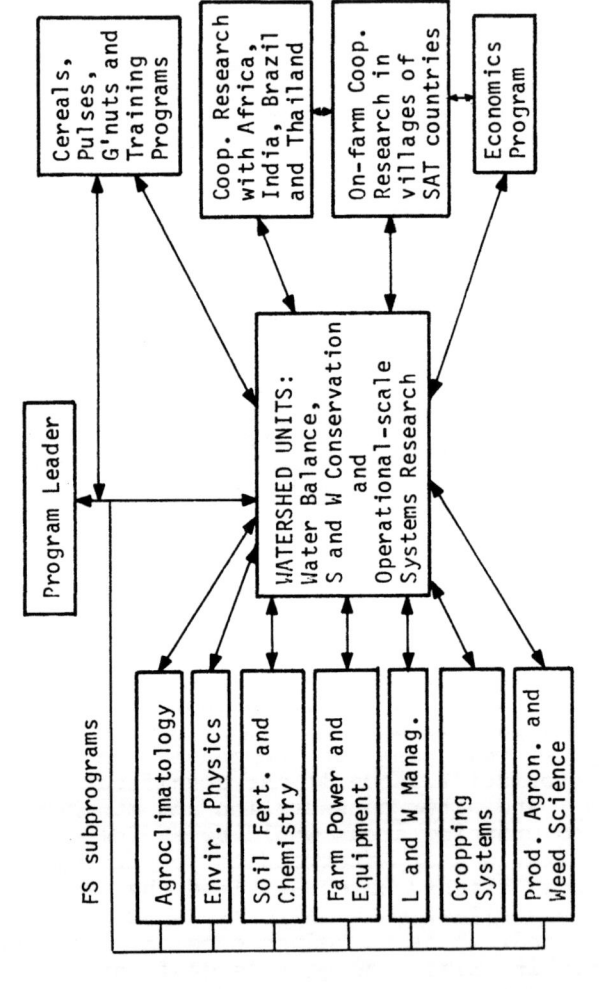

L = Land   W = Water   S = Soil

nutritious forage; and provide a storage facility (tank) to collect and store surface runoff from high-intensity storms as backstopping for rainfed agriculture.

The Watershed-Based System of Soil and Water Conservation

Since water is the first limiting natural factor in crop production in the SAT, improving the management and conservation of water and soil for increased crop production becomes the primary aim of farming systems research. In rainfed agriculture, the only water available is the rain that falls on a given area. Thus, the watershed (catchment) is the natural focus of research on water management in relation to crop production systems, resource conservation, and utilization (Krantz, 1978 and 1979).

Contour bunding, with adjustment to fit the field boundary bunds, is being routinely implemented in India on both Alfisols and Vertisols. Substantial expenditures for bund construction continue year after year even though there is no known recent research which shows a positive effect on rainfed crop production.

Contour bunding, in comparison with watershed-based resource utilization, employs distinctly different concepts of water conservation and management. In contour bunding, the excess water may flow in a concentrated manner, causing erosion between bunds. The runoff collects at the bund and is then forced to flow across the slope and out of the watershed where it is finally disposed of in roadside drains or gulleys.

In cropped watersheds cultivated in graded beds and furrows, excess water is allowed to flow through small field furrows to the grassed drainage ways and is then safely conducted to a tank and/or outlet. The velocity of flow of the water is controlled by the direction and slope of the bed-and-furrow system and runoff concentration in large overland flow is avoided. Since the 150-cm bed-and-furrow system can remain in place as a "semi-permanent" land feature, it can provide considerable protection against soil erosion on a year-round basis, even during the prolonged hot and dry noncrop season, when occasional high intensity rains occur. Broadbed furrows were established in 1975 in Alfisols and in 1976 in Vertisols. The beds have remained in place as a semi-permanent feature since that time with primary tillage as shown in Figure 2 and final bed reshaping (Fig. 3) being carried out each year.

The slope used in any soil should minimize erosion during high intensity rain, increase infiltration, provide adequate crop drainage during prolonged rains (especially on deep Vertisols), and facilitate supplemental irrigation when needed.

142

Fig. 2. Primary tillage immediately after harvest of the second crop with a left and right hand plow and a chisel or sweep in center. (This plowing concentrates organic residues in the plant zone and reforms the bed leaving a rough cloddy surface which is very receptive to pre-monsoon showers.)

143

Fig. 3. Ridger-cum-bed former being used for reshaping beds on a moist Alfisol just before planting. The semi-permanent beds were established four years ago and have been maintained in the same place with minimum tillage.

Investigations on the Bed-and-Furrow System

Systems involving graded (150 cm) beds separated by furrows which drain into grassed waterways appear to fulfill the requirements of the soil and water conservation and management listed above. The improved surface drainage function of beds and furrows compared to flat cultivation has been shown by Chowdhury and Bhatia (1971) and Krantz and Kampen (1973).

In Alfisols, the 75-cm beds were found to be unstable and cross flow and erosion were sometimes encountered, especially in slight depressional areas. This problem was overcome by the use of a 150-cm bed-and-furrow system which was started in the 1975 season. The 75-cm beds were also found to have very limited flexibility to accommodate the wide range of crops grown in the SAT. With the 150-cm beds it is possible to plant two, three, or four rows per bed at 75-, 45-, and 30-cm row spacings, respectively (Fig. 4).

In the watershed units, flat cultivation was compared with bed and furrow systems in both intercropped and sequential cropping during 1976 and 1977 (Table 1). In the deep Vertisols, the average monetary value for each of the four crops was consistently better with beds and furrows as compared with the flat system. The mean gross monetary value of the grain for the bed-and-furrow system was Rs. 650/ha greater than in the flat system. Since the average cost of the bed-and-furrow system was Rs. 74 less than that of the flat system, the net advantage of the beds and furrows over the flat system was Rs. 724. Thus, the net return was especially good with intercropping in the bed-and-furrow system on the deep Vertisol (Rs. 4,980 - 1,470 = 3,510). The gross monetary value trends were less consistent in the shallow to medium Vertisols than in the deep Vertisols and the increase of the bed over the flat system was not significant.

The beds function as "mini-bunds" at a grade which is normally less than the maximum slope of the land. Thus, when runoff occurs, its velocity is reduced and infiltration opportunity time increased. The excess water is removed in a large number of very small flows. Thus, the permanent bed-and-furrow system provides water control for *in situ* soil and water conservation throughout the year. Preliminary data at ICRISAT indicate that the optimum slope for the bed-and-furrow system is 0.3 to 0.6 percent in Alfisols and 0.4 to 0.8 percent in Vertisols. Some additional features of this system observed in operational-scale research on natural watersheds include the following:

1. Only minor earth moving (smoothing) is required.
2. No land is taken out of production.
3. The beds can remain in place as "semipermanent" features and thus no contour bunds or field bunds are necessary (Fig. 2 and 3).

Fig. 4.   Some possible row arrangements for various cropping
patterns on narrow and broad beds.

Narrow beds and furrows are adapted to 75 cm rows only

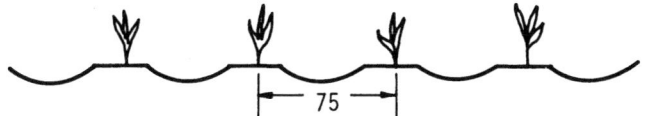

Broad beds and furrows are adapted to many row spacings

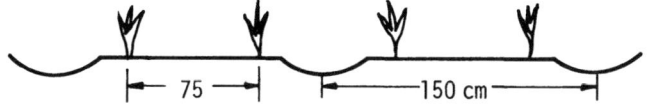

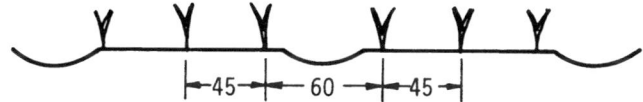

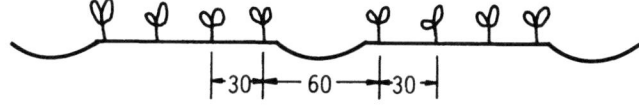

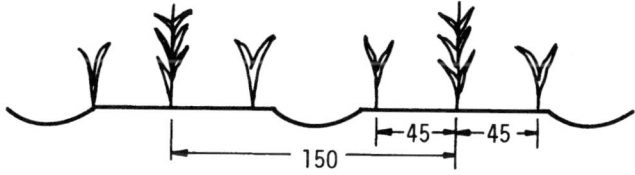

Table 1.   Mean gross monetary values of rain in flat vs. semipermanent
bed-and-furrow system on Vertisol watersheds using improved
technology in 1976 and 1977.

Water-shed no.	Land manag.	Year	Intercrop			Sequential crop			Means[*]	
			Maize Rs/ha	P.pea Rs/ha	Total Rs/ha	Maize Rs/ha	Ch.pea Rs/ha	Total Rs/ha	Both systems	Both years
			A.	Deep Vertisols						
1,2,3A	Beds	1976	2840	2080	4920	2730	950	3680	4300	
1,2,3A	Beds	1977	2270	2770	5040	2880	2400	5280	5160	
Means										4730
3B, 4B	Flat	1976	2530	1680	4210	2300	570	2870	3540	
3B, 4B	Flat	1977	2450	1810	4260	2790	2200	4980	4620	
Means										4080
LSD (.05)										280
C.V.%										9.2
			B.	Shallow to medium deep Vertisols						
7B,C,D	Beds	1976	2020	1570	3590	1970	560	2530	3060	
7B,C,D	Beds	1977	2460	1630	4090	2410	1550	3960	4030	
Means										3550
6C, 6D	Flat	1976	1960	1490	3450	1570	560	2130	2790	
6C, 6D	Flat	1977	2310	1880	4190	2290	1390	3680	3950	
Means										3370
LSD (.05)										N.S.
C.V.%										15.6

[*]The 1977-1978 costs of inputs, labor, bullock power, and depreciation of equipment
for the-bed-and furrow and flat systems were Rs. 1663 and 1737, respectively. The
Rs. 74 lower cost for the bed and furrow system was due to the smaller amount of
time required for land preparation and cultivation in the semiperminent beds and
furrows compared to the flat system. The average costs of the sequential crop and
intercrop systems were Rs.1930 and 1470, respectively. The Rs.460 higher cost in
the sequential crop system is due to the extra land preparation, seed, fertilizer,
and planting cost of the second (sequential) crop. (The cost data were supplied
by the ICRISAT Economics Program; Rs.8 = one U.S. dollar.)

4. Preliminary "shear vane" measurements indicate that soil compaction of the wide bed (plant zone) is less than under flat cultivation.

5. The surface soil of beds dries more quickly between early monsoon showers than does the surface soil on flat cultivated areas, thus facilitating the planting on beds.

6. The system can be used within the farmers' field boundaries in one of the Vertisol watersheds.

7. Soils on the beds remain friable through the cropping season. On Vertisols, primary tillage can begin immediately after harvest (Fig. 2). The beds and furrows can be maintained with minimal tillage with animal power (Fig. 3).

The Efficient Use of Animal Power with Improved Implements

The pros and cons of using animal power have been discussed by Johnston (1978) and Uzureau (1974). Research at ICRISAT indicates that it is possible to implement proper soil, water, and crop management systems using bullocks as the primary source of power for cultural operations provided that the proper machinery is available. In the semiarid tropics farm sizes are small and capital resources limited, and thus animal power is well suited to these small farms.

At least 16 to 20 hectares are usually required to make the ownership and operation of a tractor a viable proposition. Binswanger (1978) in his review of numerous tractor studies in South Asia shows that on smaller farms tractors are hired out to a much greater extent. Kline (et al., 1969) states that in northern Ghana, a holding of four to six hectares of crop land is necessary to justify a farmer's owning a pair of oxen. In contrast, Subrahmanyam and Ryan (1975) state that in India, farmers having two or three acres own a pair of bullocks. In many countries of the SAT, tractors are imported and thus foreign exchange is required for purchase of the tractor and subsequent fuel and spare parts. Bullocks or buffalos are an indigenous source of power. Ramaswamy (1978) reports that in India there is more animal power (30,000 mW equivalent) than installed electrical capacity (26,000 mW).

In countries such as India where the use of animal power has been traditional for many centuries, it is well known and understood by most farmers. While there are several hundred thousand tractors in India, most of these are concentrated in the northern irrigated areas. Subrahmanyam and Ryan (1975) using 1966 data show that in states such as Haryana and Punjab only 69 and 57 percent of the agricultural power is derived from animals. In such semiarid states as Andhra Pradesh, Karnataka, and Madhya Pradesh, 86, 89, and 96 percent of the agricultural power is derived from animals. Thus, the small farmers in SAT India practicing rainfed agriculture still rely almost exclusively on animal power.

It is often stated that animals require a large amount of grain and compete with humans for food. However, draft animals consume mainly fodder and grazing of grasslands which are often unsuited for cropping. Although grain is usually fed during the field work season, it is often possible to feed cull grains that are unsuitable for human consumption.

At ICRISAT a multipurpose animan-drawn, wheeled tool carrier is used for all cultural operations on an 80 hectare operational research area. Much of this land is double cropped. The wheeled tool carrier consists of a tool bar frame with two pneumatic tires and a beam for attaching the bullock yoke. A great variety of implements can be attached to the tool bar making it fully as versatile as a tractor. The size of the implements and depth of tillage can be adjusted to soil working conditions and the draft available from a pair of bullocks.

The wheeled tool carrier provides both horizontal and vertical precision. The horizontal precision means that implements will tract in a straight line without any effort being expended by the operator to guide or control it. Vertical precision refers to the control of depth at which an implement works which is equally important. For example, the depth at which a seed is placed is often critical to within one or two centimeters. If soil engaging tools used for tillage go too deeply, they create unnecessary and excessive draft; if the depth is too shallow, the quality of work is poor.

Where contour farming is practiced, such as in the graded bed-and-furrow system at ICRISAT, the use of a wheeled tool bar is essential to provide the stability required to keep cultivation implements in the precise line on the beds. In land preparation, preliminary results indicate that the efficiency of the wheeled tool carrier is several fold greater than that of the traditional implements. Thus with improved implements and timely operation, fewer bullocks are required and less land is required to grow the forage and grain needed to feed the animals needed for draft power.

An additional major advantage of the wheeled tool carrier is that it can also be used for transportation by placing a cart body on the chassis. In this way the farmer has added versatility and extended usage of the equipment at very little additional cost. Where hauling is a major enterprise, the chassis can be used as the front wheels of a four-wheeled unit.

Operators prefer to use a wheeled tool carrier because of the reduction in drudgery. Also, more work will be accomplished in a day if the operator can ride because his fatigue is greatly reduced and the speed at which the animals walk is not affected by the walking speed of the operator.

Water Intake and Runoff of Alfisols and Vertisols

The saturated hydraulic conductivity of the Vertisols is very low compared with the Alfisols. However, at the onset of the rainy season (when both soils are very dry), the initial infiltration rate is equally high (about 75 mm/hr) on both soils.

Thus, in spite of the low terminal hydraulic conductivity of the deep Vertisols, the water intake capacity early in the monsoon season is high due to deep cracks and the large water-retention capacity. The high initial infiltration rate is further enhanced if the soil management is such that the surface soil is rough and cloddy and is prepared in a bed-and-furrow system on a graded contour. In contrast, the initially high infiltration rate of Alfisols is often greatly reduced during the early rainy season by surface sealing caused by the impact of raindrops on the bare soil. Thus, the runoff from cropped Alfisols is usually much greater than that from cropped Vertisols (Table 2). These data are in contrast to the generally accepted statement that Vertisols have greater runoff than Alfisols (Vandersypen et al., 1972). The latter comment appears to be based on the comparative hydraulic conductivity of these soils under saturated conditions.

Under monsoon cropping in the bed-and-furrow system, the Vertisol surface dries quickly making it receptive to the next rain. The whole profile is usually near saturation only for short periods during the latter half of the season. However, during the rainy season in the flat cultivated fallow system, the Vertisol profile becomes saturated by mid-season, and runoff and erosion are thus greatly increased during the remainder of the season (Table 2).

These runoff data have great practical significance for appropriate water management on these two soils. Since Alfisols have a low water retention capacity, crops will frequently experience moisture stress during breaks in the rainy season. These can be expected to occur more than once every two or three years in many areas of the SAT. If a water storage facility (tank) is provided in a small watershed, the early runoff from Alfisols can be collected, stored, and used as a supplemental "lifesaving" irrigation until further rain comes.

In contrast, the deep Vertisols which have a greater water storage capacity and less runoff during the early rainy season rarely require supplemental irrigation for the rainy season crop. During the rainy season in each of six years at ICRISAT, high yields have been obtained on Vertisols without supplemental irrigation. In all six years the planting was made in dry soil just prior to the onset of the rainy season.

Table 2. Rainfall and runoff on a cropped Alfisol and a cropped deep Vertisol watershed with bed-and-furrow system at 0.6 percent slope and a monsoon-fallowed watershed, 1976.

		Runoff		
		Alfisol	Deep Vertisol†	
Date	Rainfall*	Cropped	Cropped	Fallow
	(mm)	(mm)	(mm)	(mm)
23 June	23	1.8	0	0.5
2 July	24	3.0	1.7	0.2
21	89	25.0	16.9	49.4
4 August	32	8.5	2.3	21.4
19	105	77.5	27.0	95.4
20	39	16.5	19.5	37.1
21	10	0	4.2	8.5
26	8	0.5	0.1	3.2
4 September	20	2.3	0.4	11.1
Ten small storms	149	5.3	0.9	11.4
Total	499	140.4	73.0	238.2

* Includes only rainfall from the 19 runoff-producing storms. The total rainfall for the monsoon season (June-October) was 679 mm.

† In 1976 the soil losses in the rainy season cropped and fallowed deep Vertisol watersheds were 0.8 and 9.2 ton/ha, respectively.

The Effect of Soil Management upon Runoff and Soil Loss

Recent results show that runoff and soil loss can be greatly reduced by improved management in deep Vertisols. In 1976, the greatest runoff was caused by a storm on August 19th, when 105 mm of rain fell. In the fallowed Vertisol, 95 mm of this rain ran off indicating the vulnerability of fallowed (bare) deep Vertisols to runoff and erosion (Table 2). The soil erosion from this storm in fallowed Vertisol and cropped Vertisol watersheds was 7.43 and 0.26 tons/ha, respectively. During 1974 to 1977, the average annual soil erosion in the traditional rainy season fallowed Vertisol and in the improved Vertisol watersheds was 5.1 and 0.6 tons/ha, respectively. The respective annual crop values were 980 and 5,090 Rs/ha. In addition to the soil loss observed at the outlet of the watershed, substantial erosion could be observed in the cultivated fallow watersheds between contour bunds.

In temperate semiarid regions with annual rainfall in the 200-mm range, fallowing during one or more years will often increase grain yields due to the large quantities of stored moisture available to the crop (Pengra, 1952). However, in the SAT high intensity rains greatly exceed the infiltration capacity of the soil and total seasonal rainfall is frequently several fold the capacity of the root zone to store water. In deep Vertisols, cultivated fallowing is practiced during the rainy season with cropping only during the post-rainy season. In India about 18 million hectares of deep Vertisols are monsoon-fallowed and post-monsoon cropped (Malone, 1974). The reasons for not cropping during the rainy season are many, including such factors as poor drainage, difficulties in tillage and weed control, and inadequate soil and crop technology (Kampen and Associates, 1974). However, the consequences of this traditional fallowing system in deep Vertisols are serious with regard to soil erosion. Jacks et al. (1955) noted that a few minutes of high intensity rainfall on some bare soils are sufficient to cause surface sealing and drastic reduction of infiltration. Ellison (1944) and Hudson (1973) pointed out the serious consequences of cultivated fallow systems on soil erosion and the critical importance of vegetative cover during high intensity storms.

Under the climatic conditions experienced at ICRISAT during its first six years of operation, the practice of cultivated fallow during the monsoon has shown no advantage in terms of moisture conservation or post-rainy season crop yields when compared with areas cropped during the rainy season.

Contour or graded bunding (terracing) has been used successfully in western countries in farms with large fields. In the SAT, field sizes are small (0.2 to 0.9 ha); bunds constructed on the contour usually would bisect the farmers' small fields. The farmer objects to

this and the soil conservation technician is forced to
"adjust" the contour bund to the field boundary. As a
result, water is impounded and the bunds are often breach-
ed by nature or by man during intense rains (Chittaranjan,
1977).

Runoff Collection and the Use of Supplemental Water

The results of supplemental irrigation to crops on
Alfisols during a 30-day drought during late August and
early September of 1974 were quite spectacular. Yields
of sorghum and maize were approximately doubled by the
application of a 5 cm irrigation. At product prices
prevailing at the time of harvest, gross rupee values of
the average increase due to the application of a 5 cm
supplemental irrigation at a critical time of growth in
two watersheds were 3,120; 2,780; 1,085; and 650 Rs/ha
for maize, sorghum, pearl millet, and sunflower, respectively.

During the 1975 rainy season, rainfall was uniformly
distributed and irrigation was not required. In the post-
rainy season, however, sorghum on deep Vertisols responded
to supplemental irrigation at the grain filling stage.
In one watershed a single 5 cm irrigation increased yields
from 2,570 to 3,570 kg/ha.

On Alfisols, tomatoes planted on beds in pearl
millet stubble yielded 12.7 tons without irrigation. In
spite of unusually heavy and late rains in October and
early November, there was a marked response to supplemental irrigation. The yields of areas receiving 0.0 cm,
2.5 cm, and 5.0 cm (in two 2.5 cm applications) of supplemental irrigation were 12.7, 17.2, and 22.2 metric
ton/ha. The yields in a flat-planted watershed were con-
siderably less due mainly to the difficulty of applying
irrigation water.

Transforming Labor into Capital

The FSR program at ICRISAT is investigating various
means of improving the natural resource base by using
labor intensive technology involving human labor and
animal power with improved implements. This activity
includes small watershed development involving graded
contour tillage for soil and water conservation; water
collection, storage, and use; drainage; and ultimately
the reforestation of eroded steep lands which are now
being cultivated. Newland (1979) points out that these
types of labor intensive projects "would have the effect
of transforming abundant labor into valuable capital."
This approach, she adds, which would enable more multiple
cropping and increased productivity, would also provide
more permanent employment for landless laborers and would
help to reduce the disparity between the landless and the
landed.

Summary

The Semi-Arid Tropics (SAT) are characterized by undependable rainfall which creates high risk and is the major cause of persistently low and unstable crop yields. Population increases have caused expanded cropping into unsuitable lands, resulting in greatly increased runoff and soil erosion. Past approaches to improved soil and water conservation have not provided the basis for substantially increased food production.

Alfisols and Vertisols are the two most abundant soil orders of the SAT. These soils, which may occur in adjacent areas, have distinctly different profile characteristics due mainly to the type and amount of clay. An understanding of these differences is essential for the development of improved management systems.

In spite of their lower saturated hydraulic conductivity, deep Vertisols, due to surface cracks, have a higher initial intake rate and less runoff in the early rainy season storms than do Alfasols. The greater early season runoff in the Alfisols provides greater opportunity for water collection and storage for supplemental irrigation during breaks in the monsoon.

The requirement for supplemental "lifesaving" irrigation during breaks in the monsoon is frequent on Alfisols and rare on deep Vertisols; crops on both soils benefit from supplemental water in the dry season.

By timely tillage of deep Vertisols during the dry season, "dry planting" of crops such as sorghum, pigeonpea, and maize just before the monsoon rains has been successful in six years of research at ICRISAT. Dry planting on Alfisols with their low water retention capacity is risky.

Based on 70 years of rainfall data at Hyderabad, the median length of growing season on the Alfisols and Vertisols was calculated at 17 and 26 weeks, respectively.

Under the traditional system of farming of the Vertisols, three-fourths or more of the rain is lost by evaporation, runoff, and drainage beyond rooting depth. With improved technology these losses can be substantially reduced and crop production greatly increased and stabilized.

Due to management problems and the lack of seedbed preparation technology, deep Vertisols are normally fallowed during the rainy season and cropped only during the post rainy season. Watersheds under rainy season fallow produced much lower crop yields and had about eight times as much erosion as did double-cropped watersheds.

With the development of improved soil, water, and crop management systems and proper selection of crops, it is possible in most years to crop most deep Vertisols during both seasons. On Alfisols, intercropping techniques and/or the availability of supplemental water

facilitates growing two crops on at least part of the
land.

The watershed based farming systems, using graded
150 cm bed-and-furrow systems at 0.4 to 0.6 percent slopes
with grassed waterways and small tanks, show potential
for reduced soil erosion, more effective rainfall use,
improved surface drainage, possibilities for supplemental
irrigation, reduced risk, and greatly increased crop
yields on Alfisols and Vertisols. Land development and
all cultural practices for all systems can be done with
bullock drawn implements.

An animal drawn wheeled toolbar used in field-scale
operational research at ICRISAT has been found to have
precision and versatility equal to that of a tractor but
at a small fraction of the cost. It can also be quickly
converted to either a two or four wheeled cart for transport purposes.

Improved animal drawn implements have been found to
be several fold more efficient for tillage operations
than traditional implements and thus fewer bullocks are
required. Riding a wheeled implement reduces human
drudgery and is more prestigious than walking behind a
wooden plow. The use of improved implements also en-
courages an integration of improved crop and livestock
farming.

References

Anonymous, 1970. A new technology for dryland farming.
    Indian Agricultural Research Institute, New Delhi,
    India. p. 5.
Binswanger, H. P. 1978. The economics of tractors in
    South Asia. Agricultural Development Council, New
    York.
Binswanger, H. P., B. A. Krantz, and S. M. Virmani. 1976.
    The role of the International Crops Research Insti-
    tute for the Semi-Arid Tropics in farming systems
    research. ICRISAT, Hyderabad, India.
Chittaranjan, S and U. S. Patnaik. 1977. Safe disposal
    of water through vegetated channels and not ponding
    against level bunds should be the approach in black
    soils. Informal seminar paper at Karnataka State
    Department of Agriculture, India.
Chowdhury, S. L. and P. C. Bhatia. 1971. Ridge planted
    kharif pulses yield despite waterlogging. In Indian
    Farming, June 1971.
Ellison, W. D. 1944. Studies on raindrop erosion. Agr.
    Engr. 25: 131-136.
Hudson, N. W. 1973. Soil conservation. B. T. Batsford
    Limited, London, England.
International Crops Research Institute for the Semi-Arid
    Tropics, ICRISAT annual reports 1973-1974, 1974-1975,
    and 1975-1976.

Jacks, G. V., W. D. Brind, and P. Smith. 1955. Mulching. Tech. Commun. Commonw. Bur. Soil Sci., 49.

Johnston, Bruce. 1978. Agricultural production potentials and small farmer strategies in sub-Saharan Africa. In Two studies of development in sub-Saharan Africa, by S. N. Acharya, and B. Johnston. World Bank staff working paper No. 300, World Bank, Washington, D. C.

Kampen, J. and Associates. 1974. Soil and water conservation and management in farming systems research for the semiarid tropics. Paper presented at the International Workshop on Farming Systems, ICRISAT, Hyderabad, India. November 1974.

Kline, C. K., D. A. G. Green, R. L. Donahue, and B. A. Stout. 1969. Agricultural mechanization in Equatorial Africa. Michigan State University, East Lansing, Michigan.

Krantz, B. A. and J. Kampen. 1973. Water management for increased crop production in the semiarid tropics. Proceeding of National Symposium on Water Resources in India and Their Utilization in Agriculture, T. K. Sarkar, editor, Water Technology Center, IARI, New Delhi, India. pp. 145-171.

Krantz, B. A. 1979. Small watershed development for increased food production. ICRISAT leaflet ICR 719-0019.

Krantz, B. A. 1980. Soil and water management for increased food production in the semiarid tropics. A Rockefeller Foundation publication of an International Conference on Integrated Crop and Livestock Production to Optimize Resource Utilization on Small Farms in Developing Countries. Bellagio, Italy. Approved as ICRISAT journal article No. 47, February 1979.

Malone, C. C. 1974. Indian agriculture: progress in production and equity. Ford Foundation, New Delhi, India.

Newland, Kathleen. 1979. How labor can become capital. Agenda--USAID, May 1979.

Pengra, R. F. 1952. Estimating crop yields at weeding time in the great plains. Agron. J. 44: 271-274.

Ramaswamy. 1978. The planning, development, and management of animal energy resources in India. Indian Institute of Management, Bangalore. Occasional paper No. 10.

Ryan, J. G. 1976. Resource inventory and economic analysis in planning agricultural development in drought prone areas. Paper presented at a training program for agricultural officers in DPAP districts, Hyderabad, India. February 9-14, 1976.

Subrahmanym, K. V. and J. G. Ryan. 1975. Livestock as a source of power in Indian agriculture: a brief review. ICRISAT, Hyderabad, India.

156

Troll, C. 1966. Seasonal climates of the earth. World
    maps of climatology, Springer Verlag, Berlin,
    Heidelberg, Germany, and New York.
Uzureau, C. 1974. Animal draft in West Africa. World
    Crops, 26(3): 112-ff.
Vandersypen et al. 1972. Handbook on hydrology.
    Government of India, New Dehli, India.

10
Farming Systems Concepts Arising from the TAC* Review and from Personal Experience

Donald L. Plucknett

Objectives

What are the objectives of farming systems research (FSR)? It was mentioned earlier that we want to raise farm income, which is one of the major objectives. Many of the talks today also have emphasized improved technology at the farm level. This, too, is very important, and I do not think that we can dismiss it. But, there are other purposes for which we can use farming systems research in a productive way for the benefit of the country. One is to learn what the farmers are doing. Partly this may be for problem identification and partly to give research direction or programs direction for the future. There is also a great need just to understand what the farmer is doing.

Ken McDermott likes to talk about farmer wisdom. I believe very much in this. We had a discussion about how wise farmers really are, and whether in some areas they really are using the best practices, or at least good practices for that environment. I think you could make a case that in a lot of areas they are using very good practices, and that until we gather and understand the knowledge they have, we really do not have the knowledge we need in that area. We must understand what they are doing and, if possible, why.

I can give you an example of that. Two years ago Dick Harwood and I were in China looking at vegetable farming systems which are probably the most complex systems in the world. Dick and I stood and scratched our heads for many days trying to figure out what was really going on in those complex fields where so many crops were being used. It is interesting that in China, the major

---

*TAC is the acronym for the Technical Advisory Committee of the Consultative Group on International Agricultural Research.

information that is being used as the basis for extension
materials is not a product of "research" *per se* at all.
Most of it has come from sending scientists and other
people down to the farm level (communes) to learn from
the farmers, analyze what they are actually doing, record
it, understand it as best they can, draw out (where it is
possible) the theory and reasons to understand it, and
then publish the information in extension materials.
These extension materials are very effective and well
illustrated.

Another example is the practice of planting crops in
the middle of the slope of the furrow rather than on top
of the ridge or the bottom of the furrow. This was observed in Egypt by a U. S. scientist who came home and
analyzed the salt concentrations across the furrow and
found that this was the point where there was the least
salinity. He said this is what the California growers
and other people should be doing. They did, and it worked here too, of course. That is an old practice which
came straight out of traditional farming systems.

I could mention yet another example from Ecuador
that I found fascinating. The Indians in the Andes use a
serpentine irrigation system which employs bunds that run
up and down hill. They are spread about 15 feet apart,
depending upon the slope. Water is run down the hill in
a serpentine system, back and forth between these bunds.
The depth and angle of the furrows and the amount of
grade of these particular loops determine the water
velocity. You can irrigate on hillsides that are tremendously steep with very little soil erosion at all and
grow all sorts of crops this way. I have never seen it
except in this area of Ecuador.

I contend that there are many things that we ought
to be finding out from traditional farming systems, and
that by itself is enough justification for farming systems research in some areas. Of course, we may want to
go farther than that for most areas. We also want to
understand the farmer well enough to work with him to
improve his system. The farmer's participation is very
important and necessary.

When I was on the World Food and Nutrition Study of
Farming Systems, we were asked to come up with recommendations on what should be done in farming systems research that would make a difference. Our committee met
and decided that we really need some work on methodology.
Rather than say, "We are going to work more on a wheat
system," or whatever, we need to do a better job of
methodology and gain a better understanding. One of the
things that we decided was that if you did some of this
work to understand the natural resources and the socio-
economic environment, followed by some on-farm studies,
you could already begin to identify some policy and other
problems without any research at all and make a difference. These problems need to be brought to the attention

of readers by saying, "Look, this is really hard on these people," or it could be something positive. I think you can find a lot of problems and situations here without having to do research. Of course, some of it would be economic research.

TAC Review

The Technical Advisory Committee (TAC) of the Consultative Group on International Agricultural Research asked three of us--John Dillon from Australia, Guy Vallaeys from France, and myself--to do a review in 1977-78. This was what they called a "stripe analysis," i. e., to look at one topic across all the international center research programs which in this case was on farming systems research programs. The reason they wanted the stripe review was that many of the donors were raising such points as: "We do not really know what these FSR programs are doing. We do not understand. We look at IRRI's[2] program and it is doing one thing. We look at ICRISAT[3] and it is doing something else. We go to IITA[4] and it does not even look like the same program as at IRRI and ICRISAT. Also, CIAT[5] has dropped its program; at the same time national programs are starting. What is it we are doing? We are putting more and more money into FSR programs, and what is it all about?

Our review team looked at farming systems research across the centers and it was very rewarding and interesting. We also looked at some national and some regional programs. I had a chance to review a little of the work at CATIE[6], and we visited the Senegal program, which is national.

One of the things that was obvious to our team was that there was really no conceptual framework that was

---

[2] IRRI is the acronym for the International Rice Research Institute (Philippines).

[3] ICRISAT is the acronym for the International Crops Research Institute for the Semi-Arid Tropics (India).

[4] IITA is the acronym for the International Institute for Tropical Agriculture (Nigeria).

[5] CIAT is the acronym for the International Center for Tropical Agriculture (Colombia).

[6] CATIE is the acronym for the Tropical Agricultural Research and Training Center (Costa Rica).

elucidated and in print for farming systems research.
There were concepts from IRRI, ICRISAT, and IITA that
were good but each program looked so different. IITA had
a heavy emphasis on soil taxonomy and land resources.
ICRISAT was placing heavy emphasis on water and water
modelling and rainfall patterns. IRRI was doing something different again. Much of this did not make sense
to some people, but we decided that there really were
good reasons why people were doing the things they were.
In part it was because of the type of staff they had, but
it was also due to the site in which they found themselves.

We could make a strong case for IITA doing land resource work in Africa, because that was one of the major
problems it faced. Its staff had to know the land resources in the humid and sub-humid tropics with which it
was working, how to classify areas as targets of opportunity for increased use which are now being used primarily
for shifting cultivation or for short-bush fallow, what
to do if sedentary agriculture was to be practiced there,
etc. There was a need then to understand the land resource first of all.

At ICRISAT you had to understand the water question,
as Bert Krantz has said, because that was the overriding
issue. When you went to IRRI, its program took direction
because it was working on rice-based systems. ICRISAT
was not focused only around one crop, because it did not
have as narrow a crop mandate. Rather, it worked with
more crops. IITA had a geographical kind of focus, and a
land type of focus, so it was working with a number of
crops that few ever understood--tropical vegetables,
fruits, and root crops.

Three Categories of Research

After a while, we began to notice some patterns and
to begin to see some unifying thoughts, i. e., concepts
of why people were doing this or that. For our own
purposes, we finally split these down into three areas.
We called them base data analysis, on-farm studies, and
research station studies. As we began to look at these,
it was quite clear why IRRI, ICRISAT, and IITA were not
doing the same things. IITA was involved in land classification and capability work. That is a base data
analysis type of activity under our classification. Base
data analysis in general requires and uses secondary data.
On-farm studies and research station studies tend to require original data. ICRISAT's program in water resources
also can be classified as base data analysis.

Research Station Studies

Now, if you take a look at the start of new farming
systems programs, by and large they begin on the research

station. What do we work on? We work on hunches, biological and technological opportunities, intuition,
guesses that sometimes turn out badly--anything. We
start at the experiment station, but soon begin wondering
why the farmers are not adopting some of our findings.
This leads us to wonder and say, "What is it the farmers
are really doing? How similar is our experimental work
to the farmers' activities?" Eventually, we end up
directly studying the farmers and the farmers' environment. As a case in point IRRI's program started on
the experiment station with Dr. Bradfield's work. Next,
he and his colleagues decided that they needed to understand the farmers better. Eventually, they had to learn
more about the farmers' land and other resources and the
climate. Then, they began working on the natural resources (base data analysis). Now, you do not have to
start any one way to be effective in farming systems
research; but you ought to start with a felt need so as
to understand better what is going on and how to improve
the farmers' systems.

On-Farm Studies

If, when doing on-farm studies, we can use secondary
data to help us identify the farms and for what purpose,
it would be a big help. For example, we might identify
some agro-climatic zones or targets of opportunity. We
heard some talk about this today. If we could use this
kind of information to help us focus our efforts a bit
better, this would be good. Some studies and farming
assistance programs might get along quite well with these
two types of activities (on-farm studies and research
station studies) and with only an occasional reliance
on base data analysis. As a matter of fact, we might
phase some of these activities where at some point we
need certain types of skills. Then, one might hire
consultants for base data analysis, as I think IRRI did
in some cases, and proceed to on-farm studies.
You can do various kinds of things in on-farm
studies. One would be initial surveys to find out what
the farmers are doing. This could be the reconnaissance
work that Peter Hildebrand was talking about this morning,
or a sort of initial look at what is happening on the
farm. Then, you might want to proceed to another type
of activity on the farm--that of on-farm trials. These
on-farm trials could be of various types, but might very
likely be researcher-managed trials or farmer-managed
trials.
Another type of on-farm trial only began to be
mentioned today, which is related to adoption questions.
For instance, how can we monitor adoption, rates of
adoption, and so forth when we are just going into an
area and must rely principally on baseline data?

If from a methodology standpoint you look around the world to see who has done a lot of work with on-farm studies, IRRI has done the most--both regarding depth of experience and methodology. We were very impressed with IRRI's on-farm studies and CATIE's on-farm work. I think that it behooves us all to try to learn as much as possible from these programs and then to try to see which methods might be most useful for national programs.

One of the concerns I have is that when national programs begin to work in farming systems research they start on the experiment station because that is the place where they are most comfortable. Most people know how to lay out a replicated trial. Most people have ideas, good or bad, that they want to test, and they can start easily on the station. It is when you start on the farm that it is really difficult. It is hard to do well.

Base Data Analysis

There is a real need to take a good look at base data analysis. How can we use secondary data better-- much better--than we have in the past? It is foolish for us to grind along in this area if we can save ourselves some time by doing a better job. Can we be more creative in defining agroclimatic zones? I am glad to see Jen Hu Chang here today because Jen Hu is one of the few agroclimatologists I know who has tried to take a look at the productivity of a particular zone from an agricultural standpoint. His work on productivity in the humid tropics is outstanding.

We can be more creative in making use of secondary data and basic information. We can use soil classification much more creatively than we ever have before. We are going to need to have people who look at natural resources from the standpoint of how these can serve systems-oriented research. If base data analysis is good, it should be used in such a way that it can help us to understand what is happening on the farm so that better use can be made of climatic, soil, and socio-economic data.

New Approaches

In addition to natural resource information, there are all sorts of anthropological questions of why people behave the way they do. Are there areas where farmers might behave somewhat alike so that you could begin to look at systems?

Zandstra believes that we can very rarely carry that kind of data load. Also, he says once we measure we must test the hypothesis that our area is homogeneous; therefore, we must have a lot of replications. This tells us if our original definition of boundaries has been erroneous.

But, there has been some very creative work in this area. For example, Allan Moore from Australia has done

some creative things with just using the soil profile
data available in everybody's filing cabinets. He has
learned how to use this information to draw soil boundaries that are helpful in narrowing our understanding of
things. I think we get back to Don Winkelmann's idea of
"non-perfectabilitarian" work. I think he is right on
this. We do not need to be so accurate that we define
everything. We can make some gross measurements that will
still be helpful. That is why I have been pushing the idea
of an ecological approach to systems work, because we
essentially are trying to understand things in a dynamic
way. I am an agronomist. I was taught to understand the
field plot, but I have come to believe the best thing we
could ever do for systems work is to throw away the field
plot. If we could get away from the plot, begin to make
measurements in the farmer's field, and get various disciplines to make these measurements—whether we are the
crop physiologist, the agronomist, the soils man, or the
crop protection person—we would understand what is really
going on in that dynamic way and we would be better off.

There are ecological ways of measuring these things
and of measuring what goes on in a dynamic environment.
An ecologist can go into a grassland and he can make
measurements that help him to understand what is going on
in that grassland. A fire can come through, an animal
can graze, lots of different things can happen, and he
still has a way of measuring in a general way what is
going on there. Not so with the field plot. As soon as you
have something missing, you lose sensitivity and accuracy
in the procedures. It seems to me we have to break out
of some of our disciplinary thinking in our methodologies.
This is one of the points I wanted to make here today.
When it comes to research, I think we can do a better job
of base data analysis. I guess I cannot give any real
suggestions on this except to say that I think we ought
to put some of our efforts toward it.

In addition, on-farm studies are tremendously important. Very few people know how to do these well. Most
of the people who do know how to do them are in this
room. Surely out of this we can come up with some suggestions for national programs so that they can do them
well, too.

Regarding research station studies where we look at
single factors or multiple factors in one crop, we know
how to do this very well. However, when we begin to mix
two crops, we are in unfamiliar territory. When two
crops are grown together, you get different harvest dates,
you get interactions, and the effects of one crop on
another. I would recommend to you some of the work that
is going on at ICRISAT where Bob Willy is doing some outstanding intercropping work. He has conducted some ele-
gant experiments which are truly helpful to us all when
we begin to mix crops. Beyond that, I do not think we
know how to do research station studies on systems them-

selves. Besides, I do not think in most cases that research stations are going to be doing systems research
anyway. Research stations are going to be doing component work--what we have called in our report component or
sub-component research. So you are essentially beginning
to break down factors for the purpose of disaggregating
them. Then you pull out those factors you can handle so
that you can look at them more closely.

One other comment on our report and I will close.
Some people have not fully understood what we were driv-
ing at in the report. One of the things we tried to do,
and I think it bears mentioning, was to write a conceptual framework for farming systems research and the
terminology that goes with it that could serve farming
systems generally. We did not restrict ourselves to
cropping systems. We tried to make it broad enough so
that it could be used for animal systems, too, so that
it would not have to be redone sometime. We tried to
make the terminology as broad as possible. You can dis-
agree with it, rewrite it any way you want to, but we put
down in our report what we believe farming systems research is in a way that would have broad, general use.

Conclusion

I throw out, in closing, one challenge to the agricultural economists. During the winter season in Egypt
about one-third of the land area at all times is planted
to berseem clover. In order for Egypt to meet its requirements for cotton, another third of the land needs to
be planted in cotton. Now what is happening? Because
berseem brings more money than cotton, the berseem is
grown longer in the spring--often stretching into summer--
which is forbidden by law. It is actually against the
law to grow berseem in summer because the cotton leaf-
worm builds up on berseem. Also, because fodder brings
more money than cotton, the period of berseem is extended
past the planting date of cotton. More farmers than not
grow cotton. Some plant a crop of napier grass to take
care of the rest of the summer, and they grow fodder
right on through the year. Each year Egypt is falling
progressively farther behind in its cotton crop because
the fodder need is greater. My challenge to the economists: we really need some data on the opportunity
costs of fodder. What are the real costs in these livestock economies, particularly in places like Egypt,
Pakistan, and parts of India where irrigated lands are
used for growing fodder thereby foregoing a cash crop?
The impact must be terrific, and there needs to be a look
at this as to both positive and negative aspects. It is
not well understood, and it seems to me it should be.
This is a farming systems problem.

Acronyms

BAPPEDA	A provincial planning agency in Indonesia
BIMAS	An Indonesian production program for lowland rice
CATIE	Centro Agronómico Tropical de Investigación y Enseñanza (Tropical Agricultural Research and Training Center), Costa Rica
CGIAR	Consultative Group on International Agricultural Research, Washington, D. C.
CID	Consortium for International Development, Tucson, Arizona
CIAT	Centro Internacional de Agricultura Tropical (International Center for Tropical Agriculture), Colombia
CIMMYT	Centro Internacional de Mejoramiento de Maíz y Trigo (International Maize and Wheat Improvement Center), Mexico
CRIA	Central Research Institute for Agriculture, Indonesia
CSU	Colorado State University, Fort Collins, Colorado
FAO	Food and Agriculture Organization of the United Nations, Rome
FS	Farming Systems
FSR	Farming Systems Research
FSR&D	Farming Systems Research and Development
HYV	High Yielding Varieties
ICRISAT	International Crops Research Institute for the Semi-Arid Tropics, India
ICTA	Instituto de Ciencia y Tecnología Agrícolas (Agricultural Science and Technology Institute), Guatemala

165

IDRC	International Development Research Centre, Canada
IICA	Instituto Interamericano de Ciencias Agrícolas (Inter-American Institute of Agricultural Sciences), Washington, D. C.
IITA	International Institute for Tropical Agriculture, Nigeria
INIAP	Instituto Nacional de Investigaciones Agropecuarias (National Institute for Agricultural Research), Ecuador
INMAS	An Indonesian production program
IRRI	International Rice Research Institute, Philippines
LDC	Less Developed Country
MRN	Ministerio de Recursos Naturales (Ministry of Natural Resources), Honduras
MSU	Michigan State University, East Lansing, Michigan
PCARR	Philippine Council for Agriculture and Resources Research, Philippines
PNIA	Programa Nacional de Investigación Agropecuaria (National Program for Agricultural Research), Honduras
ROCAP	Regional Office for Central American Program, Guatemala
TAC	Technical Advisory Committee of the Consultative Group on International Agricultural Research, Washington, D. C.
SAT	Semiarid Tropics
USAID	United Stated Agency for International Development, Washington, D. C.

Notes on the Authors and Papers

Richard R. Harwood--Is director of the Organic Gardening
and Farming Research Center, Rodale Press, Inc.,
Kutztown, Penna. He was formerly head of the Cropping Systems Program at IRRI.[1] His paper was presented at the Farming Systems Research and Development Workshop in Fort Collins, Colo., August 1-4,
1979.
David W. Norman and Elon H. Gilbert--David Norman is an
agricultural economist and professor in the Economics Department at Kansas State University. He spent
11 years in Nigeria where he was head of the Department of Agricultural Economics at Ahmadu Bello
University in Zaria. Elon Gilbert has been a Program Advisor in Agriculture for the Ford Foundation
and the Deputy Director of the Economics Institute
at the University of Colorado. Their paper was
prompted by a workshop on farming systems research
in Mali sponsored by the Institut d'Economie Rurale
and the Ford Foundation held in Bamako, Mali, November 14-19, 1976. Many of the ideas contained in
this paper are treated in greater length in Gilbert,
E. H., D. W. Norman, and F. E. Winch. 1980. *Farming
Systems Research: A Critical Appraisal.* MSU Rural Dev.
Paper No. 6, Dep. Agric. Econ., Michigan State Univ.,
East Lansing, Mich.
Donald F. Winkelmann and Edgardo R. Moscardi--Donald
Winkelmann is head of the Economics Program at
CIMMYT, headquartered in El Batan, Mexico. Edgardo
Moscardi is the Regional Economist for CIMMYT, head-
quartered in Quito, Ecuador, where he assists the
National Institute for Agricultural Research (INIAP).
Their paper was prepared for the seminar on Socio-
Economic Aspects of Agricultural Research in Devel-
oping Countries held in Santiago, Chile, May 7-11,
1979.

---

[1] See the List of Acronyms at the back of this book for
an explanation of this and other acronyms.

168

Robert D. Hart--Has recently joined Winrock International
     in Morristown, Arkansas. Before that, he was a pro-
     duction systems agronomist with the Annual Crops
     Program at CATIE in Turrialba, Costa Rica. His
     first paper was presented at an Iowa State Univer-
     sity-CATIE-IICA seminar on Agricultural Production
     Systems Research held in Turrialba, February 19,
     1979. The companion paper involving a case study in
     Honduras was presented at the Farming Systems Re-
     search and Development Workshop in Fort Collins,
     Colo., August 1-4, 1979. This paper was also re-
     produced as *One Farm System in Honduras: A Case Study,
     1980. In Activities at Turrialba.* 8:1:3-8. CATIE,
     Turrialba, Costa Rica.
Hubert G. Zandstra--Is Associate Director (Animal Sci-
     ences) with the IDRC in Vancouver, British Columbia.
     Before that, he was Head of the Cropping Systems
     Program for IRRI in Los Banos, Philippines. His
     paper was presented at the World Bank in Washington,
     D. C., on August 1, 1979 and at the Farming Systems
     Research and Development Workshop in Fort Collins,
     Colo., August 1-4, 1979.
Peter E. Hildebrand--Is a professor in the Food and Re-
     source Economics Department at the University of
     Florida at Gainesville. Earlier he was an agri-
     cultural economist with the Rockefeller Foundation
     assigned as the Coordinator for the Rural Socio-
     economics Group of ICTA in Guatemala. His paper was
     prepared for a conference on Integrated Crop and
     Animal Production to Optimize Resource Utilization
     on Small Farms in Developing Countries held at the
     Rockefeller Foundation Conference Center in
     Bellagio, Italy, October 18-23, 1978. The opinions
     expressed in the paper are those of the author and
     do not necessarily convey ICTA policy.
Jerry L. McIntosh--Is cropping systems agronomist with
     the cooperative CRIA/IRRI program in Bogor, Indone-
     sia. This paper was prepared for the Cropping Sys-
     tems Working Group meeting of the Indonesian Nation-
     al Program in Bogor, July 20-21, 1979.
Bert A. Krantz--Is Emeritus Soils Specialist at the
     University of California at Davis. Previously, he
     was Leader of the Farming Systems Program at
     ICRISAT. This paper was presented at the Farming
     Systems Research and Development Worshop in Fort
     Collins, Colo., August 1-4, 1979. He acknowledges
     the use of data from the staff of the Farming Sys-
     tems Program at ICRISAT.
Donald L. Plucknett--Is Scientific Advisor to the CGIAR
     in Washington, D. C. Before that, he was Chief of
     the Agricultural and Rural Development Division,
     Office of Technical Resources, Bureau for Asia,
     USAID in Washington, D. C. This paper is based on

his presentation at the Farming Systems Research and
Development Workshop in Fort Collins, Colorado,
August 1-4, 1979.

Index

Tansley, Arthur, 45
Technical Advisory Committee.
    See Consultative Group on
    International Agricultural
    Research, Technical Advisory
    Committee
Technological Package project, 24
Technology
  acceptance of, 100-109
  biological sciences and, 33
  component, 74-75, 89-91, 120
  cropping patterns and, 79, 82,
    83-84(Figures 3 and 4), 89-91
  development systems, 33-43,
    102-106
  economics and, 33
  environmental interactions,
    75-76
  intercropping, 8-9
  management, 82, 83-84(Figures 3
    and 4)
  returns on, 82
  utility of, 32-43
Tropical Agricultural Research
    and Training Center (CATIE)
    (Costa Rica), 159, 162

Upstream farming systems research.
    See Resource management re-
    search

Ustalfs. See Alfisols
Usterts. See Vertisols
Uzureau, C., 147

Vallaeys, Guy, 159
Vertical system interaction, 45-46
Vertisols, 136, 153
  irrigation of, 149, 150(table 2),
    152
von Bertalanffy, L., 45

Water conservation
  bed-and-furrow system, 144,
    145(Figure 4), 146(Table 1),
    147
  runoff collection, 152
  runoff data and, 149
  in semiarid tropics, 139, 141
  watershed-based system, 141,
    142-143(Figures 2 and 3), 154
Werge, Robert, 106
Willy, Bob, 163
Winkelmann, Don, 163
World Bank, 74
World Food and Nutrition Study of
    Farming Systems, 158

Zandstra, Hubert G., 162